ANIMAL RITES

ANIMAL RITES

American Culture, the Discourse of Species,
and Posthumanist Theory

Cary Wolfe

FOREWORD BY **W.J.T. MITCHELL**

THE UNIVERSITY OF CHICAGO PRESS
Chicago and London

CARY WOLFE is professor of English at the State University of New York at
Albany. He is the author, most recently, of *Critical Environments: Postmodern Theory and
the Pragmatics of the "Outside"* and the editor of *Zoontologies: The Question of the Animal.*

The University of Chicago Press, Chicago 60637
The University of Chicago Press, Ltd., London
© 2003 by The University of Chicago
All rights reserved. Published 2003
Printed in the United States of America
12 11 10 09 08 07 06 05 04 03   1 2 3 4 5

ISBN: 0-226-90513-6 (cloth)
ISBN: 0-226-90514-4 (paper)

Library of Congress Cataloging-in-Publication Data

Wolfe, Cary.
    Animal rites : American culture, the discourse of species, and posthumanist theory /
Cary Wolfe ; foreword by W.J.T. Mitchell.
        p. cm.
    Includes bibliographical references and index.
    ISBN 0-226-90513-6 (alk. paper)—ISBN 0-226-90514-4 (pbk. : alk. paper)
    1. Animal rights—Philosophy. 2. Species—Philosophy. 3. Humanism. 4. Human-
animal relationships in literature. 5. Human-animal relationships in motion pictures.
    I. Title.

HV4708 .W65 2003
179'.3—dc21

                                                                2002020411

For Sam, Pilar, Woody, Hugo, Wellston, and Oreo: family

CONTENTS

The Rights of Things

W. J. T. Mitchell

I am not an animal. I am a human being.
—*The Elephant Man* (David Lynch)

The question of animal rights tends to produce a combination of resistance and anxiety. Resistance, because acknowledging the claim that animals might have or deserve rights entails a revolution in thinking and behavior so profound that it would shake the foundations of human society. Anxiety because we suspect there is something compelling and irresistible about the concept of animal rights, at least insofar as we are all dimly aware that human life as now constituted is based on the mass slaughter of billions of animals accompanied by untold suffering. If "animal rights" means something like "humane treatment of animals" (meaning painless methods of slaughter, some degree of comfort for animals destined for slaughter, and prevention of cruelty more generally), then it is hard to see a way of mounting a principled opposition to it—unless one takes a firm Cartesian position that animals are merely machines without feelings or a purely nominal, legalistic position that "rights talk" is just not the right sort of talk when it comes to humane treatment of animals. But animal rights today means something more radical than "humane treatment"; so both the resistance and the anxiety it provokes are heightened, and a whole set of counterarguments is mustered.

The very notion of "animal rights," to begin with, seems impossible insofar as it is modeled on human rights, because the very idea of *human* rights (as the Elephant Man reminds us) is predicated on the difference between humans and animals. Rights, as political philosophers from Hobbes to Edmund Burke to Hegel have argued, cannot simply be given to anyone; they must be won through political struggle. The slave is emancipated not by virtue of his humanity but because he risks his life in resisting the master. The "rights of Englishmen," Burke contended, were not natural rights grounded in an abstract concept of the human, but the historical achievements of a specific people. The French Revolution's "rights of man" may appeal to a natural, universal foundation, but their accomplishment involved a political revolution. The "inalienable rights" enumerated in the Declaration of Independence may be "endowed by [the] creator" of human beings, but they were made real by military and political action. Animals are in no position to assert their rights, to struggle for their claims. Any rights they have, or are likely to have, will be those given to them by human beings.

And in any event (the argument continues), what good is it to fret about animal rights when *human* rights are nowhere near being established? First let us get our own house in order and create a humane world civilization that does not treat vast populations of human beings as if they were sheep to be shorn, cattle to be slaughtered, or vermin to be exterminated. Then we can talk about the rights of animals. In the meantime, an obsession with animal rights is nothing but the ultimate form of liberal guilt, the kind of self-indulgent breast-beating that encourages moralistic, sentimental posturing while doing nothing to change the lot of animals. Animals are simply the latest candidates in an endless procession of victims—women, minorities, the poor—clamoring for rights and justice, or just a modicum of decent treatment. If animals want rights (and who is to say they do?) then they will have to take a number. The queue of prior claimants is long.

And then suppose we did satisfy all those claims and the animals got their turn in the ascent to dignity, legitimacy, or whatever status is rightfully theirs? What then? Would we start with the primates, then the other mammals and "higher" animals, and work our way down to salamanders, fish, oysters, and ultimately to the primitive organisms that live in our own stomachs? Would we try to draw a line somewhere based on degrees of intelligence, sensitivity, and sociability? Would we start with the animals we happen to *like*, those that enjoy the status of domestic pets? Would this mean treating pets differently or liberating them from their dependent, ignoble status? And what about those animals that we like only in the sense that we like to eat them—fish, flesh, and fowl? Would an extension of rights to these creatures mean nothing but "humane slaughter," the present system of industrialized death purged of cruelty and unnecessary suffering? Or would it mean something more drastic—sacrificing our carnivorous desires in a wholesale conversion to vegetarianism? Would it mean the even more radical posture of the "vegan," renouncing even secondary animal products such as butter, cheese, and eggs? Are eggs deserving of rights because they are *potential* animals, and will their attainment of these rights be modeled on the "right to life" of the potential human known as the fetus, or even the fertilized human egg?

Let us suppose, finally, that all these issues have been worked out and the rights of all animals, high and low, have been established. Would that be the end? Or would it then be time to turn to the rights of fruits and vegetables? Erasmus Darwin noted long ago that "the loves of plants" are essential to their lives. Does that give them a claim to some sort of rights? Perhaps not the same level of rights accorded to animals, but some norms of decent, even loving treatment, and (most urgently) some limits on genetic intervention in their lives? One thinks here especially of the kinds of agribusiness that produce sterile patented seed strains that crowd out self-germinating varieties and turn whole human populations away from local agricultural self-sufficiency to a dependence on the food products of multinational corporations. At some point human self-interest, if not human rights, converges with the interests and rights of the plants, as the myth of the Garden of Eden suggests. C. S. Peirce went so far as to compare the human production of meaning to the life cycle of a sunflower following the sun and reproducing itself in the process. William Blake probably inspired this comparison with "Ah Sunflower!" his lyric on human desire "seeking after that sweet golden clime / Where the traveller's journey is done." All our pieties about "rootedness"—from the sacredness of place to the multiple "locations" of cultures and subject—are grounded (the pun is fully intended) in the figure of the human as plant. Is the right not to be uprooted from one's land a *human* right?

When we have attained these new, utopian forms of botany and zoology, however, and the rights of plants and animals have all been worked out in new forms of bioethics and biopolitics, there will still be work to do. At that point it will be time to take up the rights of *things*, of inanimate objects. This may sound like a peculiar or whimsical notion, but in fact the rights of things are already much better established than those of plants or animals, and have been for a very long time. Whole classes of objects—works of art, religious icons, valuable commodities, private fetish objects, and public totems—already have special status. The old ethical conundrum about rescuing a Rembrandt or an infant from a burning building makes sense only in a culture that already believes some objects have a strong claim to human protection, care, and loving attention. Some objects (idols, notably) even demand human sacrifice. And these beliefs are invariably grounded in some notion that inanimate objects are capable of possessing life or (as Walter Benjamin put it) "aura." Even a humble object like a child's beloved teddy bear is seen as having the right to some form of respect and indulgence, if only for the sake of the child's feelings.

The rights of things are so well established, in fact, that it might make sense to work up from them toward the animals, rather than downward from the human. A person, after all, is from one point of view just another thing. Even "the king is a thing," as Hamlet noted, a "bare forked animal" that has big ideas about itself. The humblest material object, a mere wooden table, as Marx noted, "is changed into something transcendent" when "it steps forth as a commodity": "It stands on its head, and evolves out if it wooden brain grotesque ideas, far more wonderful than 'table-turning' ever was." Small wonder that, in the era of globalized commerce, there is far more freedom of movement for inanimate things than for human, animal, or vegetable bodies.

The great virtue of Cary Wolfe's *Animal Rites* is that it makes it possible to begin thinking through these questions without succumbing to the anxiety and resistance that so often accompany them. Wolfe works through a series of philosophers—Wittgenstein, Cavell, Lyotard, Deleuze and Guatarri, Lévinas, and Derrida—who have radically reshaped the traditional view of "the" animal as a straightforward antithesis and counterpart to "the" human. These theoretical resources are then woven through a series of brilliant readings of modern literary and popular texts—from Hannibal Lecter's silenced lambs to Hemingway's sacrificial bulls to the dangerous, intelligent primates of Michael Crichton's *Congo*. One glimpses in Wolfe's readings the possible emergence of a new, postmodern bestiary. Do we not live in the age of sensitive apes and talking ants, neurotic Batmen and leather-loving Catwomen, animated frogs and singing trout, purring television sets and dinosaurs with family values?

Reified images of "the" animal, Wolfe argues, produce equally reified images of "the" human. The reduction of the complex plurality of animals to a singular generality underwrites the poverty of a humanism that thinks it has grounded itself in a human essence, a stable species identity to be secured by its contrast with animality. Heidegger's human hand versus the animal's claw, Freud's human eye versus the animal's nose, the Enlightenment's human rationality versus the animal's mechanical reflexes—all these tropes of difference are (like the Elephant Man's cry for human recognition) understandable and inevitable efforts to define and affirm the species identity of human beings. But the claim to humanity and human rights will never succeed until it has reckoned with the irreducible plurality and otherness of nonhuman or posthuman life forms, including those that (like ourselves) wear a human face.

Wolfe's book is not, then, a straightforward effort to find a philosophical grounding for animal rights as conceived, say, by Peter Singer. His point is rather to make it clear why we have to think beyond utilitarian justifications of "the rights of animals" by showing that they can be modeled only on the "rights of man" and that the rights of man are in turn based in the lack of rights in the animal. Wolfe is much more interested in reopening the question of ethics and humanism (or posthumanism) than in making arguments for animal rights as conceived at present. His book asks, in other words, what it would mean to take the question of animal rights seriously in philosophy, as a deep problem for reflection, not just as a ready-made political or ideological position ripe for action.

Animal Rites must be read in the context of a very widespread outpouring of thought on questions of culture and nature, the human sciences and biology. The question of the animal is just one component in a rethinking of a whole set of nonhuman entities that seem to take on organic, lifelike, or "autopoietic" characteristics—intelligent machines, of course, but also systems and swarms, viruses and coevolutionary organisms, corpses, corpora, and corporations, images and works of art. There is, in short, a new kind of vitalism and animism in the air, a new interest in Nature with a capital *N*. We live, as the philosopher Michael Thompson has put it, in the age of "the fetishism of DNA and the secret thereof." The philosophy of life has returned with a vengeance in the age of biogenetic engineering and bioterrorism. If Walter Benjamin were around, he might call it "the age of biocybernetic reproduction"—that is, not merely the "digital" or "information" age, but the era when the sciences of life and the technologies of computation have attained a new level of dialectical intensity, when the contradictions between "sciences of control" (cybernetics) and eruption of the uncontrollable (the biosphere, typified by computer viruses) are rampant features of everyday life.

So when Wolfe talks about the animal, he is not thinking only of chimps and whether they speak to us. It is the nonlinguistic that matters just as much as the capabilities of the higher animals—the silence, the stare, the gesture, the reflex. It is also the inhumanity of language, the arts, and social forms as such, their evolution as autopoietic systems that elude the control of their supposed "creators." Perhaps we need a new term to designate the hybrid creatures that we must learn to think of, a "humanimal" form predicated on the refusal of the human/animal binary. (I am undeterred by the unwelcome news that this word has already been coined, as Cary Wolfe informs me, "in the title of a really cheesy and blessedly short-lived TV show back in the seventies or eighties—a sort of variation on the Hulk theme.") The humanimal makes sense as a "hulk"—a wrecked structure or vessel, an object of care and salvage, not simple ownership or instrumentality; as an object of duties and obligations before it is a subject of rights; as a valued "end in itself" rather than a tool, an abject "thing" or found object. The humanimal is also, obviously, the animal as myself and my kin. When Wolfe talks about the animal in positivist science, then, he is neither adopting nor rejecting some metaphysics or epistemology. His strategy is a relentlessly close reading of a dialectical tradition, an immanent critique of the trope of the animal in the history of philosophy, and in contemporary popular culture, as the figure that is not merely below or beside "the human" but actively constitutive of the human. In a time when "identities" of race, gender, and sexuality have dominated so much of our thinking, Wolfe raises the even deeper, more intractable question of species identity. "Speciesism" is ritually invoked in the denigration of others as animals while evoking a prejudice that is so deep and "natural" that we can scarcely imagine human life without it. The very idea of speciesism, then, requires some conception of "the posthuman," an idea that makes sense, obviously, only in its dialectical relation with the long and unfinished reflection on species being that goes by name of humanism.

Animal Rites, as the title suggests, strategically circumvents the current impasse over animal *rights* by focusing on the *rituals* we construct around the figures of animals and "the animate"—our narratives of brutality and cannibalism, monstrosity and normativity, our metaphors of animation and the anima or soul. The most central ritual is, of course, the whole process of thinking out—of *imagining*—what we are as human or posthuman beings, which means what we have been and might become as well. Wolfe's book opens onto a host of interesting topics: the resurgence of totemism (the mutual mapping of human identities and natural objects) and the question of sacrificial ritual; the obsession of popular culture in our time with figures of animality, animation, artificial life forms, cloning, and biotechnology; and above all, the fundamental issue of biopower, biotechnology, and human rights in a time when so many human beings are treated as animals or worse. He has reconstructed an entire dialogue in modern philosophy that bears on all the fundamental positions and arguments. The result is a rare and valuable work of intellectual synthesis that many readers will find indispensable as they struggle to understand the new forms of life (and imitations of life) that are emerging around us today.

ACKNOWLEDGMENTS

Chapter 1 originally appeared in slightly different form in *EBR: Electronic Book Review*, no. 4, in a special issue titled "Critical Ecologies," coedited by Joseph Tabbi and Cary Wolfe, http://www.electronic-bookreview.com/ebr4/ebr4.htm, and also in *Diacritics* 28 (summer 1998): 21–40.

Chapter 3 originally appeared in *Boundary* 2 22 (fall 1995): 141–70.

The second half of chapter 4 (on *The Garden of Eden*) originally appeared in *Boundary* 2 29 (spring 2002): 223–47.

The first half of the introduction and all of chapter 5 originally appeared in somewhat different form in *Arizona Quarterly* 55 (summer 1999): 115–53.

I gratefully acknowledge permission from these publications to reproduce my work here.

Introduction

I want to begin by suggesting that much of what we call cultural studies situates itself squarely, if only implicitly, on what looks to me more and more like a fundamental repression that underlies most ethical and political discourse: repressing the question of nonhuman subjectivity, taking it for granted that the subject is always already human. This means, to put a finer point on it, that debates in the humanities and social sciences between well-intentioned critics of racism, (hetero)sexism, classism, and all other -isms that are the stock-in-trade of cultural studies almost always remain locked within an unexamined framework of *speciesism*. This framework, like its cognates, involves systematic discrimination against an other based solely on a generic characteristic—in this case, species.[1] In the light of developments in cognitive science, ethology, and other fields over the past twenty years, however, it seems clear that there is no longer any good reason to take it for granted that the theoretical, ethical, and political question of the subject is automatically coterminous with the species distinction between *Homo sapiens* and everything else.[2]

That my assertion might seem rather rash or even quaintly lunatic fringe to most scholars and critics in the humanities and social sciences only confirms my contention: most of us remain humanists to the core, even as we claim for our work an epistemological break with humanism itself. This might seem like a harsh verdict, except that the rest of United States culture has long since gotten the point about animals that is just beginning to dawn on our critical practice. Over the past several years *Time, Newsweek,* and *U.S. News and World Report* have all run multiple cover stories on new developments in cognitive ethology that seem to demonstrate more or less conclusively that the humanist habit of making even the *possibility* of subjectivity coterminous with the species barrier is deeply problematic, if not clearly untenable.[3] And PBS and cable television—most recently in the big-budget PBS series on "the animal mind" hosted by *Nature* executive producer George Page—have made standard fare out of one study after another convincingly demonstrating that the traditionally distinctive marks of the human (first it was possession of a soul, then "reason," then tool use, then tool *making*, then altruism, then language, then the production of linguistic *novelty*, and so on) flourish quite reliably beyond the species barrier.[4]

These developments, and their implications for our critical practice, have been registered largely, if at all, in the "literature and science" wing of cultural studies. As Donna Haraway puts it in perhaps the central theoretical statement of this recently established field, her famous "Cyborg Manifesto,"

> By the late twentieth century in United States scientific culture, the boundary between human and animal is thoroughly breached. The last beachheads of uniqueness have been polluted, if not turned into amusement parks—language, tool use, social behavior, mental events. Nothing really convincingly settles the separation of human and animal. . . . Movements for animal rights are not irrational denials of human uniqueness; they are clear-sighted recognition of connection across the discredited breach of nature and culture.[5]

Now my point here is not to harangue you about animal rights, but rather to point up that current critical practice, for all its innovation and progressive ethical and political agendas, takes for granted and reproduces a rather traditional version of what I will call the discourse of species—a discourse that, in turn, reproduces the *institution* of species*ism* (a point I will return to in a moment).

There is no better known and more powerful embodiment of that discourse, perhaps, than Freud's *Civilization and Its Discontents*. There the origin of humans is located in an act of "organic repression" whereby they begin to walk upright and rise above life on the ground among blood and feces. These formerly exercised a sexually exciting effect but now, with "the diminution of the olfactory stimuli," they seem disgusting, leading in turn to what Freud calls a "cultural trend toward cleanliness" and creating the "sexual repression" that results in "the founding of the family and so to the threshold of human civilization." All of this is accompanied by a shift of privilege in the sensorium from smell to sight, the nose to the eye, whose relative separation from the physical environment thus paves the way for the ascendancy of sight as the sense associated with the aesthetic and with contemplative distance and sensibility.[6] Freud's account, however, is riven with a fundamental antinomy around which it circulates like a bad conscience; for there the human being, who becomes human only through an act of "organic repression," has to *already* know, *before* it is human, that the organic is repulsive and needs to be repressed. And so Freud's "human" is caught in a chain of infinite supplementarity, as Jacques Derrida would put it, that can never come to rest at an origin forming a break with animality. This means, of course, that the figure of the human in Freud, despite itself, is constituted by difference at the origin. Or to put it in post-Darwinian rather than poststructuralist terms, the subject of humanism is constituted by a temporal and evolutionary stratification or asynchronicity in which supposedly "animalistic" or "primitive" determinations inherited from our evolutionary past—our boundedness to circadian rhythms, say, or the various physiological frailties that foreground the body as physically determined by a fundamentally ahuman universe of interactions ruled by the laws of physics, chemistry, and genetics—coexist uneasily in a second-order relation of relations, which the phantasmic "human" surfs or manages with varying degrees of success or difficulty.

Freud's valorization of the human who sees at the expense of the animal who smells is sustained (even if transvalued) in the figure of vision that runs from Sartre's discourse on the look in *Being and Nothingness* through Foucault's anatomy of panopticism in *Discipline and Punish*. This critical genealogy tells us that the figure of vision is indeed ineluctably tied to the specifically human, with the look in Sartre serving to objectify the subject and foreclose his freedom and the panoptical gaze in Foucault signaling power's omnipresence. By these lights, it is indeed tempting to abandon the figure of vision altogether. But I am sympathetic with attempts, such as Haraway's, to reorient it toward what she calls "situated knowledges" and away from its traditional phallic associations with "a leap out of the marked body and into a conquering gaze from nowhere," a gaze with "the power to see and not be seen, to represent while escaping representation."[7] Here again, confronting the problem of nonhuman others seems especially important. For if the look purchases the transcendence of the human only at the expense of repressing the other senses (and more broadly the material and the bodily with which they are traditionally associated), then one way to recast the figure of vision (and therefore the figure of the human with which it is ineluctably associated) is to resituate it as only one sense among many in a more general—and not necessarily human—bodily sensorium.

As Thomas Nagel long ago realized in framing his famous essay "What Is It Like to Be a Bat?" these phenomenological differences make the problem of the animal other a privileged site for exploring the philosophical challenges of difference and otherness more generally.[8] In her wonderful (if

sometimes frustrating) book *Adam's Task*, Vicki Hearne—a master horse
and dog trainer as well as a poet and a student of philosophy—provides two
useful examples of such difference: the dog's sense of smell and the horse's
sense of touch. As Hearne points out (following Stanley Cavell's early
work), "What skepticism largely broods about is whether or not we can believe our eyes. The other senses are mostly ancillary; we do not know how
we might go about either doubting or believing our noses." But "for dogs,
scenting is believing. Dogs' noses are to ours as a map of the surface of our
brains is to a map of the surface of an egg."[9] And so, as you sit in your garden with your dog, he sees what you see, but "what he *believes* are the scents
of the garden behind us," the cat moving slowly through it, the bird hopping
about and hunting for insects, and so on. "We can show *that* Fido is alert to
the kitty, but not *how*, for our picture-making modes of thought interfere
too easily with falsifyingly literal representations of the cat and the garden
and their modes of being hidden from or revealed to us" (80–81).

Similarly, the kinesthetic sensibility of horses is so exquisite that when
they are handled by an inexperienced rider, "every muscle twitch of the rider
will be like a loud symphony to the horse, but it will be a newfangled sort of
symphony, one that calls into question the whole idea of symphonies, and
the horse will not only not know what it means, s/he will be unable to know
whether it has any meaning or not" (108). And thus both horse and rider
find themselves squarely within the frame of what Cavell calls the "skeptical terror of the independent existence of other minds" in which both parties, as Hearne puts it, "know for sure about the other . . . that each is a creature with an independent existence, an independent consciousness and thus
the ability to think and take action in a way that might not be welcome
(meaningful or creature-enhancing) to the other." More important—and
this is crucial for properly decentering the human and the visual from its
privileged place as the transcendental signifier to which all other phenomenological differences are referred for meaning—"the asymmetry in
their situations is that the horse cannot escape knowledge of a certain sort
of the rider, albeit a knowledge that mostly makes no sense, and the rider
cannot escape knowing that the horse knows the rider in ways the rider cannot fathom" (109). As Hearne puts it, if the horse could speak, she might say,
"I still don't know people, but I can't help but fathom them" (109).

As Cavell's early work suggests, the traditional humanist subject finds this
prospect of the animal other's knowing us in ways *we* cannot know and master *simply unnerving*. And in response to that "skeptical terror," we have mobilized a whole array of prophylactics: the reinscription of the animal other
within the domesticated economy of the pet, which, as Gilles Deleuze and

Félix Guattari have argued, is essentially Oedipal and narcissistic;[10] the treatment of animals (familiar since Descartes) as mere unfeeling brutes, as stimulus-response mechanisms, or more recently, as genetically programmed routines and subroutines; or the demonization of the animal as the monster or mysterious "outsider," the figure who, as Cavell puts it, "allegorizes the escape from human nature" by "obeying his nature as he always does, must"—as the one who can't really ever be a subject at all. Musing on the folklore that dogs can "smell" fear and danger, Cavell observes that "it is important that we do not regard the dog as honest; merely as without decision in the matter" (qtd. in Hearne, 215). Cavell sums all this up in a remarkable letter quoted by Hearne in *Adam's Task:*

> There is something specific about our unwillingness to let our knowledge come to an end with respect to horses, with respect to what they know of us. . . . The unwillingness . . . is to make room for their capacity to feel our presence incomparably beyond our ability to feel theirs. . . .
>
> The horse, as it stands, is a rebuke to our unreadiness to be understood, our will to remain obscure. . . . And the more beautiful the horse's stance, the more painful the rebuke. Theirs is our best picture of a readiness to understand. Our stand, our stance, is of denial. . . . We feel our refusals are unrevealed because we keep, we think, our fences invisible. But the horse takes cognizance of them, who does not care about invisibility. (115)

Are *we* ready, Cavell asks us, to "understand" the animal—"underknow" her and thereby stand "under," not above her—by surrendering the dream of mastery troped as vision? Can *we* handle the skeptical terror of "letting our knowledge come to an end"? In posing these questions, Cavell underscores that our stance toward the animal is an index for how we stand in a field of otherness and difference generally, and in some ways it is the most reliable index, the "hardest case" of our readiness to be vulnerable to other knowledges in our embodiment of our own, an embodiment that arrives at the site of the other before we do, as our scent reaches the dog's nose before we round the corner, telling a story we can never wholly script to a present we have not yet reached.

The sorts of questions Cavell poses give a glimpse of what it is like to query anew the question of the animal and the discourse of species that makes it available, but to say "discourse" here is tell only part of the story. One might safely argue at the present moment that cultural studies and "theory" are engaged in addressing a social, technological, and cultural

context that is now thoroughly posthuman if not quite posthumanist, insofar as the "human" is inextricably entwined as never before in material,
technological, and informational networks of which it is not the master,
and of which it is indeed in some radical sense "merely" the product. As
Bruno Latour asks in calling for a "Parliament of Things," "Where are the
Mouniers of machines, the Lévinases of animals, the Ricoeurs of facts? . . .
The human, as we now understand, cannot be grasped and saved unless
that other part of itself, the share of things, is restored to it."[11] On the one
hand, then, the question of the animal is embedded within the larger context of posthumanist theory generally, in which the ethical and theoretical
problems of nonhuman subjectivities need not be limited to the form of the
animal alone (as our science fiction writers have dramatized time and
again). On the other hand, the animal possesses a specificity as the object
of both discursive and institutional practices, one that gives it particular
power and durability in relation to other discourses of otherness. For the
figure of the "animal" in the West (unlike, say, the robot or the cyborg) is
part of a cultural and literary history stretching back at least to Plato and
the Old Testament, reminding us that the animal has always been especially, frightfully nearby, always lying in wait at the very heart of the constitutive disavowals and self-constructing narratives enacted by that fantasy figure called "the human."

It is this pervasiveness of the discourse of species that has made the *institution* of speciesism fundamental (as Georges Bataille, Jacques Derrida,
René Girard, and others have reminded us) to the formation of Western
subjectivity and sociality as such, an institution that relies on the tacit agreement that the full transcendence of the "human" requires the sacrifice of the
"animal" and the animalistic, which in turn makes possible a symbolic economy in which we can engage in what Derrida will call a "noncriminal
putting to death" of other *humans* as well by marking *them* as animal.[12] The
"discourse" of my title sits, theoretically and methodologically, at the intersection of "figure" and "institution," the former oriented more toward relatively mobile and ductile systems of language and signification, the latter toward highly specific modes and practices of materialization in the social
sphere. And to broach the question of the "institution" of speciesism—as
Derrida has recently done with particular force—is to insist that we pay attention to the asymmetrical material effects of these discourses on particular groups. Just as the discourse of sexism affects women disproportionately
(even though it theoretically may be applied to any social other of whatever
gender), so the violent effects of the discourse of speciesism fall overwhelmingly, in institutional terms, on nonhuman animals.

The effective power of the discourse of species when applied to social others of whatever sort relies, then, on a prior taking for granted of the *institution* of speciesism—that is, of the ethical acceptability of the systematic "noncriminal putting to death" of animals based solely on their species. And because the discourse of speciesism, once anchored in this material, institutional base, can be used to mark *any* social other, we need to understand that the ethical and philosophical urgency of confronting the institution of speciesism and crafting a posthumanist theory of the subject *has nothing to do with whether you like animals.* We all, human and nonhuman alike, have a stake in the discourse and institution of speciesism; it is by no means limited to its overwhelmingly direct and disproportionate effects on animals. Indeed, as Gayatri Spivak puts it, "The great doctrines of identity of the ethical universal, in terms of which liberalism thought out its ethical programmes, played history false, because the identity was disengaged in terms of who was and who was not human. That's why all of these projects, the justification of slavery, as well as the justification of Christianization, seemed to be alright; because, after all, these people had not graduated into humanhood, as it were."[13]

A similar point, in terms that will be even more familiar to students of American culture, is made in Toni Morrison's eloquent meditation *Playing in the Dark: Whiteness and the Literary Imagination.* She argues that the hallmarks of the individualist imagination in the founding of United States culture—"autonomy, authority, newness and difference, absolute power"—are all "made possible by, and shaped by, activated by a complex awareness and employment of a constituted Africanism," which in turn has as its material condition of possibility the white man's "absolute power over the lives of others" in the fact of slavery.[14] My point here, however (and it is one I will press in my discussion of her reading of Hemingway's *The Garden of Eden*), is to take Morrison very seriously at her word—and then some. For what does it mean when the aspiration of *human* freedom, extended to all, regardless of race or class or gender, has as its material condition of possibility absolute control over the lives of *nonhuman* others? If our work is characterized in no small part by its duty to be socially responsive to the "new social movements" (civil rights, feminism, gay and lesbian rights, and so on), then how must our work itself change when the other to which it tries to do justice is no longer human?

It is understandable, of course, that traditionally marginalized peoples would be skeptical about calls by academic intellectuals to surrender the humanist model of subjectivity, with all its privileges, at just the historical moment when they are poised to "graduate" into it. But the larger point I stress

here is that as long as this humanist and speciesist *structure* of subjectiviza-
tion remains intact, and as long as it is institutionally taken for granted that
it is all right to systematically exploit and kill nonhuman animals simply because of their species, then the humanist discourse of species will always be
available for use by some humans against other humans as well, to countenance violence against the social other of *whatever* species—or gender, or
race, or class, or sexual difference. That point has been made graphically in
texts like Carol Adams's *The Sexual Politics of Meat*, which, despite its problems, demonstrates that the humanist discourse of species not only makes
possible the systematic killing of many billions of animals a year for food,
product testing, and research but also provides a ready-made symbolic
economy that overdetermines the representation of women, by transcoding
the *edible* bodies of animals and the *sexualized* bodies of women within an
overarching "logic of domination"—all compressed in what Derrida's recent work calls "*carno*phallogocentrism."[15]

For these reasons, the reopening of the ethical question of the animal is
an event whose importance is named but not really captured by the term
"animal rights." Indeed, as we will see in the opening chapter, one of the central ironies of animal rights philosophy is that its philosophical frame remains essentially humanist in its most important philosophers (utilitarianism in Peter Singer, neo-Kantianism in Tom Regan), thus effacing the very
difference of the animal other that it sought to respect.[16] In this, of course,
animal rights philosophy is not alone in its readiness to resort to a liberal humanism it seems to undermine in its attempt to extend the sphere of ethical
and political consideration—an approach that links the question of the animal other rather directly to other investigations in contemporary cultural
studies that focus on difference, identity, and subjectivity.

As we will see in the opening chapter, for thinking about the animal, the
liberal philosophical tradition was promising in its principled emptying of
the category of the subject. It insisted that subjectivity—and with it freedom—no longer depended on any single identifiable attribute, such as
membership in a certain race, gender, or class. And from there it was but one
short step for animal rights philosophy to insist that species too should be
set aside, that membership in a given species should have no bearing on
freedom and rights. The problem, of course, is that while the category of the
subject was *formally* empty in the liberal tradition, it remained *materially* full
of asymmetries and inequalities in the social sphere, so that theorizing about
the subject as "nothing in particular" could easily look like just another sign
of the very privilege and mobility enjoyed by those who were quite locatable
indeed on the social ladder—namely, at the top.

It is in response to what we might call this self-serving abstraction of the subject of freedom that much of the work in what is now known, for better or worse, as cultural studies and identity politics arose to reassert the social and material "location" (to use Homi Bhabha's term) or "standpoint" (to use an older vocabulary still) of the subject. The problem with *this* mode of critique, as we will see with Bhabha's work in chapter 5, is that it often reinscribes the very humanism it appears to unsettle, so that the formerly "abstract" subject of liberal humanism, though now indeed socially marked and locatable, is nonetheless "marked" by a very familiar repertoire, one that constitutes its own repression—its own "sacrifice," to use the characterization of both Derrida and Bataille—of the question of the animal and, more broadly still, of the nonhuman. As Jean-François Lyotard puts it, such a maneuver "hurries" and "crushes" everything he means by the terms "heterogeneity, dissensus, event, thing": "the unharmonizable."[17] In this light, the point of thinking with renewed rigor about the animal is to *disengage* the question of a properly postmodern pluralism from the concept of the human with which progressive political and ethical agendas have traditionally been associated. And it is to do so, moreover, *precisely by taking seriously* pluralism's call for attention to embodiment, to the specific materiality and multiplicity of the subject—not so much for the pragmatic reason of addressing more adequately our imbrication in the webworks of Latour's "Parliament of Things" (the environment, from the bacterial to the ecosystemic, our various technical and electronic prosthesis, and so on), but rather for the theoretical reason that the "human," we now know, is not now, and never was, itself.[18]

...

That issue is squarely on the table in my discussion in chapter 1 of Luc Ferry's much discussed *New Ecological Order*, which provides a textbook example of contemporary humanist philosophy's attempt to address the challenge of ecology and animal rights. Here I attempt to elucidate the contributions we may salvage from the humanist tradition for thinking through these questions—contributions we may carry with us for the rest of the book. Moreover, the humanist frame helps bring into focus how radical environmentalism and animal rights, though often associated with each other, emerge as distinct problematics, each with its own philosophical mortgages. Here the idea of "animal rights" retains an especially fitful relation to the humanist tradition, growing as it does out of the liberal justice tradition of moral thought but at the same time threatening the humanism at that tradition's very core. After a detailed critique of Ferry's discussion of the relation between radical ecology and liberal democracy, this chapter provides a extensive overview of animal rights philosophy as expounded by its two leading figures, Tom Regan and Peter Singer. While Ferry's ethical antinaturalism is a salutary corrective to certain tendencies in animal rights philosophy—he reminds us quite rightly, for example, that ethics is about concepts and not objects—his essentially Kantian position nonetheless runs aground on his inability to maintain the parallelism, absolutely crucial to humanism, between the doublets freedom/necessity and human/animal. Through a series of ever more desperate gambits, Ferry's humanism struggles to maintain that species difference coincides with ethical difference, whereas animal rights philosophy attempts to confront the fact that differences in degree between the human and the animal with regard to the freedom/necessity doublet (think here, for example, of Jane Goodall's work on the great apes) cannot be coherently maintained as differences in kind.

In its formulation of the subject of ethics, however, animal rights philosophy also runs aground, on the very same humanism it seeks to combat. For this reason, in chapter 2 I set out from Wittgenstein's famous aphorism in *Philosophical Investigations* ("If a lion could talk, we could not understand him") to undertake a major reassessment of how distinctly different figures in contemporary philosophy and theory have thought about the question of language in relation to the difference between human and animal—a question that has never been more pressing in light of recent scientific work on animals, language, and cognition. Here, however, my aim is not to investigate those claims but rather to examine the theoretical conditions of possibility under which those claims might matter, might *have a claim* on us. In this light, it is not simply "a matter," as Jacques Derrida puts it, "of 'giving speech back' to animals," but rather of rethinking the relation between language, ethics, and species itself.[19]

To that end, I begin the chapter with an examination of the extraordinarily searching work on the relation between language and phenomenology, of which we have already had a taste, in the philosophical lineage that runs from Wittgenstein through Stanley Cavell and Vicki Hearne. No one, perhaps, has thought in a more fine-grained way about the relation between "forms of language" and "forms of life" (to use Wittgenstein's terms) than these thinkers, and this project reaches a fascinating terminus in Hearne's writings on training horses and dogs, and in her attempts to understand how creatures with vastly different phenomenologies can build something like a common language and thus share a universe—a moral universe at that. In Cavell and Hearne, however, the suppleness of their phenomenological investigations is undercut by a humanism that manifests itself in an essentially unreconstructed contractarian notion of ethics we will have already seen critiqued by Regan in chapter 1, one that is inadequate to the very ethical implications toward which their work on language and phenomenology so compellingly points us.

In an attempt to see if a Wittgensteinian orientation toward questions of language and ethics can be rerouted in a more productive way, I turn to the work of Jean-François Lyotard—not least because Lyotard has made extensive use of the Wittgensteinian notion of the "language game" in relation to questions of justice and ethics and has, throughout his career, been conspicuously interested in what he calls "the inhuman." With detailed attention to *The Postmodern Condition, The Differend*, and *Just Gaming*, I attempt to show that even though Lyotard is committed to a postanthropocentric concept of the subject, he nevertheless (like Cavell) remains bound to humanism in the domain of ethics, and this is so (again like Cavell) because of a certain powerful relation to Kant. The same holds true, in the end, of a very different figure I discuss alongside both Lyotard and Derrida. Emmanuel Lévinas, like Lyotard, remains within the Kantian view that we have only "indirect duties" to animals because the animal "has no face": it cannot "universalize its maxim" and so cannot be an other to which, as Lévinas puts it, we are held hostage in the imperative "Thou shalt not kill."

Lévinas provides an important point of contact between the work of Lyotard and that of Jacques Derrida, whose recent investigations of the question of the animal I examine in some detail. Derrida's work on the animal, it seems to me, provides the most promising framework among the figures discussed for bridging the ethical and epistemological dimensions of issues that occupy me in this book. This is so for a quite specific reason: that Derrida's notion of ethics (unlike Lyotard's) and his articulation of the concepts of *écriture* and the trace (unlike Lévinas's) do not in principle exclude the question of the animal and even point in a strong sense toward the general problematic of the "inhuman" nature (to borrow Geoffrey Bennington's characterization) of communication, both human and nonhuman. For that reason, in the last section of the chapter I shift into a very different theoretical register—that of contemporary science and systems theory—and deploy the work of Gregory Bateson (on mammalian communication) and Humberto Maturana and Francisco Varela (on the evolutionary emergence of "linguistic domains") to flesh out the cross-species possibilities of what Derrida calls "the trace beyond the human." I thus attempt to give at least a snapshot of a posthumanist and transdisciplinary theory of the relation between the species, ethics, and language, conceived in its exteriority and materiality.

The third chapter intensifies the investigation of the "sacrificial" symbolic economy of "carnophallogocentrism" that structures the very concept
of the subject under modernity as diagnosed by Derrida and pays particular
attention to the homologies—but also the differences—between the discourses of species, gender, and sexual difference so conspicuously at work in
Jonathan Demme's film *The Silence of the Lambs*. It does so, however, by a
sustained engagement with psychoanalytic theory, which has served since
Freud as something like a privileged site for thinking through the coimplication of humanity and animality, simply because the questions of the body,
sexuality, the drives, and so on are always so close at hand in any psychoanalytic discussion of the question of the subject. This is so whether we understand psychoanalysis primarily as a theory and method of reading a symbolic or signifying field ordered by what we might think of as the "strange
attractors" of the body and sexuality or more strictly and canonically as a
claim about the somehow inescapable hold of the biological and sexual fact
on us all—or, as is the case here, the attempted fusion of the two in post-
Lacanian analysis.

For Clarice Starling (Jody Foster), cross-species identification lies at the
heart of an irrevocable trauma for which the law of culture, the law of the father (here, advancement in the FBI) is felt to be compensatory. In Demme's
film (as borne out by the body of critical attention the work has attracted)
this is especially difficult to see at first because of the nearly overwhelming
presence of the discourses of gender and sexuality that structure the entire
narrative. These include not only, of course, the drama of sexual identity of
the serial killer "Buffalo Bill" and the violence it generates, but also the apparently progressive story of Starling's "poor girl makes good" ascendancy
in the macho world of the FBI. That, however, turns out to be what Jonathan
Elmer and I call an ideological feint that, through the filmic conventions of
suspense and horror, misdirects our attention away from the more fundamental problem of species discourse, sacrifice, and law that anchors the film.
It is this redirection that makes Demme's film particularly beguiling (and
particularly brilliant), and it depends in large part on that other narrative
that parallels this first, the one that steals the show: the story of Hannibal
Lecter (Anthony Hopkins), "Hannibal the Cannibal," an impossible figure
who at once deftly deploys and flouts the sacrificial logic at the heart of
Western culture. Lecter is a horrific figure who is both analyst and monster,
an embodiment of both the law of culture and its repressed "Thing" (to use
Lacan's language), one who short-circuits the all-important difference between subject and substance and in so doing "outs" the sacrificial regime
that structures and anchors our modernity by pursuing it with unflinching

rigor. Lecter's message—and it is offered to us in several different registers, including cannibalism, incest, and sadoanalysis—is not "I eat animals and not, therefore, humans," but rather "I eat animals *and*, therefore, humans."

Via the character of Lecter and his relationship with Starling, Demme's film masterfully mobilizes and at the same time evacuates psychoanalytic discourse. And it does so, moreover, in a mode specific to postmodernity, for as Slavoj Žižek has pointed out, in modernism we repudiate animality and the primal forces unleashed (and necessarily repressed, if we believe Freud's *Civilization and Its Discontents*) at the margins of the *socius*, whereas post-modernism is characterized by a more ambivalent relation to the animal, the monster, and the Thing. In postmodernity, as Žižek puts it, "we abjure and disown the Thing, yet it exerts an irresistible attraction on us"—a fact neatly indexed by the film's altogether playful ending, where we are almost sad to see Lecter go (in silk suit and silly hairpiece), happy to know he is about to wreak his revenge on the odious Dr. Chilton: "I'm having an old friend for dinner."

The fourth chapter of *Animal Rites* undertakes a rereading of the work of a major modernist writer, Ernest Hemingway, in light of the problem of species discourse, with an eye toward demonstrating that the current tendency to view modernism as the more or less retrograde forerunner of our own more progressive attitudes about race, gender, or species is self-flattering and (at best) only half right. While some recent studies of the period have moved in this direction, they have usually been content to treat the problem of species discourse under a more general rubric: primitivism (as in Marianna Torgovnick's *Gone Primitive*), gender (Nancy Comley and Robert Scholes's *Hemingway's Genders*), race (Toni Morrison's *Playing in the Dark*), or ethnicity (Walter Benn Michaels's *Our America*). I want to insist, however, on the irreducibility of this problem, one that allows the discursive site of species to serve as an indispensable "off site" in the texts I will be examining, where adjacent problems of race or gender may operate or be operated by the discourse of species. Here species discourse will sometimes be deployed to "solve" questions of race, gender, and sexuality that the text generates but (as if it is realized too late) cannot answer in those terms. At other times it will serve instead to generate and keep open those very possibilities of difference.

This deployment is especially striking in Hemingway's most important early and late novels, *The Sun Also Rises* (1926) and the unfinished text *The Garden of Eden* (published in 1986). Recent critics such as Mark Spilka and Rose Marie Burwell have noticed Hemingway's intense interest in cross-gender identification and transgressive play with gender codes. What has not been understood is that the discourse of species operates in Hemingway's great early and late novels in diametrically opposed ways. In the early novel, the central instance of the institution of speciesism is, of course, the practice of bullfighting and the culture of *afición* it makes available, which serves the crucial function of refixing the codes of gender identity that Hemingway (in a characteristically modernist move) ironizes in the novel—most obviously, of course, in the sexual impotence of the hero, Jake Barnes. In the bullfight, the category of the masculine undergoes a radical bifurcation between the "cultural" and the "natural," so that the central "gender trouble" (to use Judith Butler's phrase) of the novel—that the impotent aficionado Jake is at once the most and least masculine of characters—is rearticulated in terms of the cultural identity made available through the compensatory domination of animal by human. Hence the question What is a man? gets rewritten as if it had always been What is the difference between a human and an animal? Moreover, at a stroke, this rewriting of gender in terms of species also vaccinates the homosocial community of the aficionados and the other forms of male bonding in the novel against any taint of unwanted homosexual connotation by cordoning off (to put it telegraphically) the "homo" from the "sexual."

These problems are handled rather differently, and with opposed ethical implications, in *The Garden of Eden*, which—despite the hairy-chested persona Hemingway cultivated in the thirties and forties—contains one of the most damning accounts of hunting ever written, as the young writer David Bourne condemns his father as a ruthless "friend killer" who commits an unpardonable act of violence against a noble animal. As with Clarice Starling's originary trauma in *The Silence of the Lambs*, at the origin of David's generalized sense of isolation and mistrust as an adult—a sense that drives him to write and to identify with his wife Catherine's androgynous search for new forms of identity—is the trauma of animal sacrifice. And here the crucial difference in the formal dynamics of cross-species versus cross-gender identification in the novel (a difference that turns on a certain reductive mimeticism that attends the site of gender performativity) is central to understanding that the most radical sight of alterity in this text, its species discourse, is locatable but finally unreadable by the model of psychoanalysis alone. This leads me, in turn, to continue the querying of the psychoanalytic model begun in the previous chapter, particularly on whether the concept of "identification" is adequate for making sense of cross-species relations.

What makes this novel even more interesting and ambitious is the way it intertwines this complicated interplay of gender and species with the discourse of race as it presents Catherine and David's increasingly adventurous experimentation with gender identity as a form of "getting dark." Toni Morrison has rightly drawn our attention to this trope's flirtation with an essentially racist "Africanist" discourse and its familiar stereotypes of the "primitive" other. But in overleaping the specificity of the discourse of species in the novel—and in this she is not alone, of course—Morrison's reading fails to see that these cross-racial identifications remain critical rather than merely stereotypical as long as they are anchored by the cross-species identification that leads David to reject the violence of his father and identify with Catherine. "Getting dark," that is, is a way of rebelling against, rather than siding with, the father's code and its Africanism, but only if the condemnation of his animal sacrifice that anchors this rebellion is kept in view. Hence Hemingway's novel, though it clearly partakes of the discursive Africanism remarked by Morrison, also pulls the rug out from under that very discourse by destabilizing what Etienne Balibar calls the "anthropological universals," subtended by the institution of speciesism, that racist discourse depends on.

In the fifth chapter of *Animal Rites* I also engage the specifically postmodern staging of species difference in relation to the question of colonial identity that haunts *The Garden of Eden*, but this time by discovering in Michael Crichton's novel *Congo* a deeper logic of neocolonialism that organizes the novel's species discourse, one that moves in the opposite direction from *Eden*'s use of the question of the animal to reopen rather than foreclose questions of identity and alterity. Crichton's novel—with its central character Amy, the linguistically prodigious gorilla raised in a university language lab—seems to offer an empirically informed and altogether up-to-date questioning of humanism's habit of limiting the issue of subjectivity to the human alone. As the novel unfolds, however, it becomes clear that *Congo*'s ostentatious transgression of the species barrier only reorganizes itself into a second-order resorting according to a deeper logic, whereby Amy and her wild gorilla counterparts find themselves on different sides of the species barrier according to their positions in the neocolonial project (as do the multinational ERTS expedition members and the cannibalistic Kigani tribe). Crichton's fictional universe thus seemingly offers us the most "progressive" discourse of species we have seen thus far in these pages, but only to quarantine it within a familiar "strategy of containment" (to use Fredric Jameson's phrase) in which species distinctions become problems not of ethics and ontology, but merely of management in a posthuman—and at the same time ultrahumanist—context.

That the conspicuously "postmodern" is not always genuinely posthumanist is the focus of the conclusion, which explores this crucial distinction on the site of ethics itself. A renewed attention to questions of ethics in the context of postmodernism seems counterintuitive at best, perhaps, not least because of the intensification under postmodernity of the familiar dilemma of immanence versus transcendence, or what Kant distinguished, in a different register, as the difference between the descriptive and the prescriptive, between "pure" and "practical" reason. To wit: If postmodernism seems once and for all to cast permanent epistemological suspicion on any form of foundational or representational thought, then at the same stroke all such claims to a secure basis for judgment that escapes social and historical contingency—all such claims to transcendence—are declared out of bounds, with the result that there is no escape from the immanence of judgment as it is constituted by any given society or community. And if that is the case, then the space of the outside and the other, on which ethics seems to depend for its ability to appeal to the difference between immanence (what our fleeting social standards say is good) and transcendence (the deeper question of justice against which such codes should be measured), seems to be foreclosed.

One of the more ambitious attempts to reconjugate this apparently self-defeating relation of the postmodern and ethical is Zygmunt Bauman's *Post-modern Ethics*, a patient, searching, and wide-ranging investigation arguing that it is precisely the absence of such foundations that keeps ethical questions alive. Bauman contends, squarely in line with Lévinas, that it is the "unaccountability" of the ethical relationship—the very fact that it exceeds all reason, all contractual and reciprocal obligation—that *makes* it ethical. I am obligated to the other—"hostage" to the other, as Lévinas puts it— *regardless* of the other's obligation to me. But while such a formulation might seem to be good news for thinking about the ethical question of the animal other, it only reinforces the very humanism it seems to subvert (again, squarely in line with Lévinas). It does so not only by assuming that the subjects of the ethical relationship are always already human, but also by placing the ethical relationship beyond all epistemological questioning— "prior" to it, to use Lévinasian language.

In so doing—and here the difference between Lévinas and Derrida discussed in chapter 2 is worth revisiting—the alterity of the other is once again captured and hypostatized (as "man") rather than left open (to the possibility of the nonhuman other), so that the "incalculable" essence of the ethical relationship turns out to be not so incalculable after all. For these reasons I argue, drawing on the epistemological innovations of systems theorist Niklas Luhmann, that the only way to think about the ethical relation with the nonhuman other that supposedly comes "before" the social and the epistemological is precisely *through* theory itself. Rather than freezing and reontologizing the difference between reason and its other (all its others), I argue that the other-than-human resides at the very core of the human itself, not as the untouched, ethical antidote to reason but as part of reason itself—the "trace" that inhabits it, to use Derrida's term. By thus keeping open the incalculability of the difference between reason/the human and its other/the nonhuman (animal), we may begin to approach the ethical question of nonhuman animals not as the other-than-human but as the *infrahuman,* not as the primitive and pure other we rush to embrace as a way to cure our own existential malaise, but as part of us, *of* us—and nowhere more forcefully than when reason, "theory," reveals "us" to be very different creatures from who we thought "we" were.

PART *One*

CHAPTER 1

Old Orders for New

Ecology, Animal Rights, and the Poverty of Humanism

Early in *The New Ecological Order*, French philosopher Luc Ferry characterizes the allure and the danger of ecology in the post-modern moment. What separates it from various other issues in the intellectual and political field, he writes, is that

> it can call itself a true "world vision," whereas the decline of political utopias, but also the parcelization of knowledge and the growing "jargonization" of individual scientific disciplines, seemed to forever prohibit any plan for the globalization of thought. . . . At a time when ethical guide marks are more than ever floating and undetermined, it allows the unhoped-for promise of rootedness to form, an objective rootedness, certain of a new moral ideal.[1]

As we shall see, for Ferry—a staunch liberal humanist in the Kantian if not quite Cartesian tradition—this vision conceals a danger to which contemporary European intellectuals are especially sensitive: not holism, or even moralism exactly, but that far more charged and historically freighted thing, *totalitarianism*. Ferry's concern is that such "world visions," incarnated in contemporary environmentalism, ecofeminism, and animal rights, threaten "our entire democratic culture," which "since the French Revolution, has been marked, for basic philosophical reasons, by the glorification of *uprootedness*, or *innovation*" (xxi). Ferry's thesis—it becomes explicit in his comparison of environmental legislation under the Third Reich with tenets of deep ecology in the book's second section—is that movements like these have come to occupy the space left open by the passing of the political imaginaries of fascism and communism, so that denunciations of liberalism (and its corollary in political praxis, reformism) may now be unmasked for what they are: critiques "in the name of *nostalgia*, or, on the contrary, in that of *hope:* either the nostalgia for a lost past, for national identity flouted by the culture of rootlessness, or revolutionary hope in a radiant future, in a classless and free society" (xxvi). To which Ferry responds—literally—"Grow up!" Late in the book, he tells us that we must follow through on "the adult development of the secular and democratic universe" (137) by rejecting totalizing revolutionary visions of the sort purveyed by radical environmentalism and adhering instead to liberal reformism, "the only position consistent with leaving the world of childhood" (138).

Ferry is certainly right to draw our attention to the often uncritical nostalgia and romantic holism of some varieties of environmental thought—problems that have been noted by critics from points on the map very different politically from Ferry's avowed liberal humanism. And it is certainly understandable, given the historical context, that he would join a long list of other European intellectuals in pointing out the manifold dangers to democratic society of totalizing moral schemes—dangers often represented for liberal intellectuals like Ferry by the rise of the Greens in European politics.[2] We do well to remember, too, that for European intellectuals like Ferry, liberalism retains, for understandable historical reasons, a viability and a promise toward which many American intellectuals are skeptical or even jaded. European intellectuals, conditioned by the experience of fascist authoritarianism and the strong but problematic presence historically of the Communist Party in social and intellectual life, may find in liberalism a refreshing and indeed radical democratic openness and dynamism. On the other hand, American intellectuals, conditioned to the absence of any other major political contenders, have long since grown accustomed to liberalism as the name for that "end of ideology" position that, as Fredric Jameson puts it, "can function *more* effectively after its own death as an ideology, realizing itself in its most traditional form as a commitment to the market system that has become sheer common sense and no longer a political program."[3]

But in defending democratic difference, everything hinges, of course, on precisely how such terms are framed and how difference is articulated—an index of which often may be found in how its imagined opponents are painted. Here, as we shall see, Ferry's text gives us early and ample pause, not least in its impoverished notions of "democracy" and the "human." As for the latter, Ferry wholly disengages the "human" from problems of class power and from the determinative force of both discourse (conceived not merely as rhetoric but also in the stronger Foucauldian sense I have already touched on in the introduction) and psychoanalytic investment. Similarly, Ferry's notion of "democracy" is extraordinarily thin because it is completely uncoupled (despite some gestures to the contrary very late in the book) from the problem of *capitalism* as liberal democracy's de facto economic embodiment. Given the well-known importance of both class and race in contemporary environmentalism—in debates about "environmental racism," for example, or the disproportionate exposure to toxic waste and environmental degradation borne by the poor, not to mention reminders of how middle class, and how white, the contemporary environmental and animal rights movements are—this is surprising and disabling for one as eager as Ferry is to defend the heritage of "democracy."

All this suggests that Ferry's critique of radical environmentalism remains locked within a liberal humanism that renders it impossible to make good on the desire for difference and heterogeneity that his aversion to holism expresses. It is not simply that he adheres to a definition of the liberal "human" as a wholly negative (that is, empty) sort of being open to "infinite" experimentation. *That* case, for what is sometimes called the "moral perfectionism" of the distinctly human, has been made, and made better, by Stanley Cavell and others (and not at all coincidentally, I think, in *dialogue* with postmodern theory rather than in the dismissal or misconstrual of it that we find in Ferry).[4] It is rather that the figure of "the human" in Ferry's liberal humanism turns out to be not so open ended and contentless after all but is instead "sovereign and untroubled," as Foucault once characterized him, "a subject that is either transcendental in relation to the field of events or runs in empty sameness throughout the course of history."[5] He is the one who can master discourse without being mastered by it, the one who is able to step outside into a space of pure, transparent reflection, the very systems and material structures in which he is supposedly ineluctably embedded. These include, of course, the laissez-faire capitalism that liberal humanism wants to pretend has no important bearing on the political equality that liberalism's call for "democracy" says it desires.

Though he devotes considerable space to discussions of animal rights philosophy (at least the version promulgated by Peter Singer's *Animal Liberation*) and, to a lesser extent, ecological feminism, the bête noire of Ferry's book is clearly deep ecology. Invented, if you will, by Norwegian philosopher Arne Naess, formalized and codified by Naess and American philosophers Bill Devall and George Sessions, and more recently adapted by the European Greens, deep ecology proposes a fundamental change, from anthropocentric to "biocentric," in how we view the relation of *Homo sapiens* to the rest of the biosphere. An eclectic blend (to put it mildly) of ideas drawn from Heidegger, Buddhism, Robinson Jeffers, and many other sources, the fundamental principles of deep ecology are nevertheless relatively easy to state. They have been formalized by Sessions and Naess in eight basic and often-quoted tenets:

1. The well-being and flourishing of human and non-human Life on earth have value in themselves (synonyms: intrinsic value, inherent value). These values are independent of the usefulness of the non-human world for human purposes.

2. Richness and diversity of life forms contribute to the realization of these values and are also values in themselves.

3. Humans have no right to reduce this richness and diversity except to satisfy vital needs.

4. The flourishing of human life and cultures is compatible with a substantial decrease of the human population. The flourishing of non-human life requires such a decrease.

5. Present human interference with the non-human world is excessive, and the situation is rapidly worsening.

6. Policies must therefore be changed. These policies affect basic economic, technological, and ideological structures. The resulting state of affairs will be deeply different from the present.

7. The ideological change is mainly that of appreciating life quality (dwelling in situations of inherent value) rather than adhering to an increasingly higher standard of living. There will be a profound awareness of the difference between big and great.

8. Those who subscribe to the foregoing points have an obligation directly or indirectly to try to implement the necessary changes. (Qtd. in Ferry, 67–68)

There is much to remark on here, but the philosophical platform of deep ecology may be boiled down to this: the ultimate good is not harmony with nature, or even holism per se, but rather something much more specific: *biodiversity*. Once this is recognized, we must affirm the *inherent value* of all forms of life that contribute to this ultimate good, and we must actively oppose all actions and processes by human beings and their societies that compromise these values.

The appeal of deep ecology and its demand that we recognize the inherent value of the biosphere and conduct ourselves accordingly is understandable for all sorts of scientific, ethical, historical, and political reasons. As Gregory Bateson points out in his influential collection *Steps to an Ecology of Mind*, "The last hundred years have demonstrated empirically that if an organism or aggregate of organisms sets to work with a focus on its own survival and thinks that that is the way to select its adaptive moves, its 'progress' ends up with a destroyed environment. If the organism ends up destroying its environment, it has in fact destroyed itself." The Darwinian paradigm of "organism versus environment" and "survival of the fittest" must be revised, Bateson argues, to read "organism-in-its-environment."[6] Rather than seeing these two terms as naming different and hierarchically related ontological orders—in which "environment" is merely a fungible resource for the self-realization and self-perpetuation of the organism—we do better, as good ethics and as good *science*, Bateson argues, to understand that both are components of a larger network or system of relations in which negative feedback is crucial to maintaining systemic balance. The Enlightenment face of Darwinism would tell us that the organism that most successfully exploits and maximizes its environmental resources is the one that wins, the one that lives to pass on its genes. But "if this is your estimate of your relation to nature *and you have an advanced technology*," Bateson tells us, "your likelihood of survival will be that of a snowball in hell" (462).

This is a central theme, of course, in the literature of deep ecology. As Hans Jonas, one of its leading European exponents, writes, "The promise of modern technology has reversed itself into a threat. . . . The subjugation of nature with a view toward man's happiness has brought about, by the disproportion of its success, which now extends to the nature of man himself, the greatest challenge for the human that his own needs have ever entailed" (qtd. in Ferry, 76–77). Ferry's first impulse—in a rhetorical strategy endemic to the book—is to dismiss such critiques as "a return of the old science fiction myths," the latest instance of Frankenstein and the Sorcerer's Apprentice, where "we have a reversal by which the creature becomes its master's master" (77). But such concerns have been raised, of course, by scores of critics and philosophers who are as far from "deep" as Ferry himself (Martin Heidegger, Kenneth Burke, Theodor Adorno, and Jeremy Rifkin come to mind, to name four rather different examples).[7] Indeed, one need not be captivated by Frankenstein scenarios to acknowledge that practices such as the current headlong rush into genetic engineering of plants and animals entail all sorts of unforeseeable consequences, inhumane practices, and potential biological disasters. Similarly, it is hard to disagree with Jonas that there is currently no way—legally, economically, or politically—to effectively control such practices, a problem made even more acute, as Ferry recognizes, by the considerable economic incentives involved (Ferry, 77). For these reasons and others—increased deregulation of business and industry, the weakening or proposed abolition of government agencies devoted to environmental protection, the increased pressure to "privatize" public lands and allow wilderness preserves and refuges to be exploited for their resources, unabashed attempts to severely weaken or abolish landmark environmental legislation such as the Endangered Species Act, the increased momentum and publicity of the "wise use" and "property rights" movements, and so on—deep ecologists and others have called for greater government activism and more forceful use of state power to regulate and direct the effects of human society and technology on the environment.

But the devil, as they say, is in the details. As several critics have pointed out, the philosophical platform of deep ecology is marked not only by eclecticism but also by incoherence. As Tim Luke has noted, if all life forms are given equal inherent value, and if biodiversity as such is an ultimate good, then we face any number of rather vexed ethical questions. Luke asks, "Will we allow anthrax or cholera microbes to attain self-realization in wiping out sheep herds or human kindergartens? Will we continue to deny salmonella or botulism micro-organisms their equal rights when we process the dead carcasses of animals and plants that we eat?"[8] And if biodiversity as such is an ultimate good, then by definition, "rare species and endangered individuals in rare species . . . are more valuable than more abundant species and individuals," creating scenarios like the following: "If one was caught in a spring brushfire a deep ecologist would be bound ethically to save a California condor hatchling over a human child, because the former—given its rarity—is much more valuable" (87). Moreover, the deep ecology platform—for all its talk of "hard" biocentrism and its "no compromise" posture—is fundamentally compromised, in its own terms, by its "vital needs" or "mutual predation" loophole, which reveals that deep ecology reverts in the final instance to a "soft anthropocentrism," one that thus remains tied to the very Enlightenment schema it means to overturn (83).

But this point of divergence between critiques from the Left, such as Luke's, and those from the Right, such as Ferry's, also provides an important point of contact, one that brings to light an essential confusion of categories at the heart of deep ecology's ethical project. For both Luke and Ferry, deep ecology attributes human qualities, and gives at least somewhat human status, to the nonhuman realm of nature. As Luke points out, "Nature here speaks of virtues and freedoms that are those of sovereign individuals," and the modern liberal paradigm of subjectivity "is not so much overcome as much as it is made into an equal entitlement and guaranteed to everything in the ecosphere, knowing all along"—as Luke reminds us in an important addendum—"that humans still have the best crack at enjoying these benefits" (84–85). For Ferry, a similar categorical mistake lies at the heart of deep ecology—but Ferry defends the very Enlightenment humanist tradition that Luke's Marxist-informed perspective would critique. There are two related but distinct points here. First, Ferry is quite right to point out that the deep ecologists, "imagining that good is inscribed within the very being of things," forget "that *all valorization, including that of nature, is the deed of man and that, consequently, all normative ethic is in some sense humanist and anthropocentrist*" (131). I will return in a moment, and in the next chapter, to *what* this "in some sense" means, exactly—and why this claim does not function in the way Ferry imagines. But for now we should surely agree with Ferry that "it is still *the ideas*, and not the object as such, that are the basis for value judgements *which only men are capable of formulating: ethical, political or legal* ends never 'reside in nature'" (141).

The last part of Ferry's assertion leads us to the second important point about ethics that he raises against the deep ecologists: that "basing ethics in biology is insufficient, for the fact that nature 'says yes to life' does not imply an *ethical* necessity that men act in favor of its preservation" (81 n. 28). That is to say, *even if* an ethics could be derived from nature, the deep ecology version of it forgets that though life forms as a rule pursue self-preservation, it is possible "to have values other than those of self-preservation, to prefer a life that is short but good, for example, to one that is long and boring" (85). Self-preservation cannot be a moral imperative because—well, it is not an imperative. Ferry here confronts the problem of naturalism in ethics that we will see explored with considerable subtlety in the next chapter in Derrida's reading of Heidegger's attempt to theorize the *Geschlecht* (or species being) of the human while at the same time avoiding biologism and its pernicious uses. Ferry is right to point out the real danger of believing that "Nature *in* itself contains certain objectives, certain goals . . . independent of our opinions and our subjective decrees" (86). Trying to derive ethical principles from empirical knowledge of biology may seem innocent enough, but when you start "with the idea that, in principle, each individual possesses a 'healthy and identical' human nature," you may be "gradually led to associate all supposedly deviant practices with pathology": "evil is confused with abnormality: one has to be crazy to smoke, not to love nature as one *should*, and so on" (89).

Keeping in mind the appalling historical track record of using the supposedly self-evident moral imperatives of nature to countenance social and political practice, Ferry is right to be worried about claims by Jonas and other deep ecologists that serious ecological reform "seems impossible, or at least infeasible . . . within the framework of a democratic society," that "we must have recourse to force," to "State constraint" (77). And he is right to put pressure on the deep ecologists on the issue of population control: "When we get to the point of arguing that the ideal number of humans, *from the point of view of nonhumans*, would be 500 million (James Lovelock) or 100 million (Arne Naess), I would like to know how one plans to realize this highly philanthropic objective" (75). "No serious democrat will argue," Ferry writes, "with the idea that it is necessary, if not to limit the deployment of technology, at least to *control and direct it.*" But "the idea that this control must occur at the price of democracy itself is an additional step which deep ecologists, propelled as they are by a hatred of humanism and of Western civilization . . . almost never hesitate to take" (78).

These are, it seems to me, the most forceful and worthwhile points that Ferry makes in his critique of deep ecology. But the passage I just quoted, while it provides a snapshot of some of what is right about Ferry's position, also suggests much of what is wrong about it—not least his reliance on overly simple oppositions between concepts like "democracy" and "totalitarianism." It is easy enough, for example, to point out in response to his critique of deep ecologists' calls for state constraint that there are all sorts of areas of social and political life in which government involvement and state power are exercised at the expense of "pure" democracy (itself, need it be said, a fiction). If we conceive of the environmental crisis as a problem fundamentally of national and international security, as well we might, then how could Ferry object to uses of state power to address the environmental crisis that are no greater than those indulged in by the military-industrial or intelligence complexes as a fact of everyday life in liberal democratic society? For short of a full-bore endorsement of anarchism (which, from Ferry, seems unlikely!) the problem is *how* and *when* such uses of state power are justified, not simply a matter of equating, as Ferry does, deep ecologists' calls for the use of state power with a "hatred of humanism and of western civilization."

In this light, it will come as no surprise that Ferry's use of the term "democracy" systematically represses the *economic* context of capitalism that historically accompanies it. Chantal Mouffe's critique of a similar problem in Richard Rorty's liberalism would apply doubly to Ferry. The problem with both is

> [the] identification of the political project of modernity with a vague concept of "liberalism" which includes both capitalism and democracy. . . . If one fails to draw a distinction between democracy and liberalism, between political and economic liberalism; if, as Rorty does, one conflates all these notions under the term *liberalism*; then one is driven, under the pretext of defending modernity, to a pure and simple apology for the "institutions and practices of the rich North Atlantic democracies."[9]

The question of where "democracy" begins and ends, in other words, is not limited to the use of state power but depends at least as much on the uneven distribution of *economic* power in capitalist democratic society. Deploying a well-worn strategy of liberal intellectuals, Ferry is eager to condemn all those who stray from liberal humanism as fanatical, totalitarian, haters of modernity, of Western civilization, of humanity—in a word, as ideological.[10] But as any Marxist would be quick to point out, he relies on a strategically vague notion of democracy that—though supposedly without any positive content—is ideological through and through.

Ferry makes a few hollow gestures very late in the book toward recognizing how the economic fact of capitalism might complicate and compromise the abstract democracy he promotes, but they are just that—hollow gestures. He observes, for example, that "no one can remain indifferent to a questioning of the liberal logic of production and consumption" (128). But when it comes to any serious examination of the relation between the de facto economic form of liberal humanism and environmental devastation, we find the same sort of laissez-faire posture we witnessed earlier in his concept of democracy. Ferry's proposed program for the "reformist ecologist" countenanced by his liberal humanism—as opposed to the revolutionary "deep" ecologists of the "new ecological order"—argues that "ecology ultimately blends into the market, which naturally adapts to new consumer demands . . . clean industry is developing by leaps and bounds, creating competition among companies to obtain the 'green' label. The supreme pardon? Perhaps. But why take offense if it allows us both to advance the cause of environmental ethics and include it within a democratic framework?" (145–46). Instead of a rigorous examination of the relation between democracy, capitalism, and environmental protection, we find in Ferry the same sort of superficial faith (this time in the "free market") that he finds intolerable in the "zealots" of deep ecology. For as Arran Gare points out in *Postmodernism and the Environmental Crisis*, the much ballyhooed use of market mechanisms to control environmental degradation—such as issuing pollution "shares" to restrict emissions to tolerable levels and then allowing companies that exceed these levels to buy more shares from companies that meet the standards—may have been enthusiastically embraced by industries and business organizations, but in practice they have failed to protect the environment. "In particular," Gare writes,

> it has been found that utilizing the market through the issuing of tradeable pollution rights, tradeable rights to exploit resources, has not achieved any significant reduction in pollution, diminution in the

rate of exploitation of mineral reserves or reduction in the rate of destruction of resources. . . . The only legislation that has had real effect
has been absolute bans on the exploitation of animal species or the use
or production of particular types of material. . . . That such measures
should have failed is a reflection on the limitations and defects of the
market as a device for regulating economic, let alone social and political, activity.[11]

In light of Ferry's thin and indeed nearly legalistic concepts of democracy
and liberal humanism, his apparent lack of knowledge about the nitty-gritty
details of environmental reform (and as we shall see, about areas relevant to
animal rights such as cognitive ethology), and because of his readiness to
rely on simplistic oppositions between ideal types such as "democracy" and
"totalitarianism," moments such as the following cannot help but come off
as pompous and a bit comical, even if we agree in spirit with Ferry's point:

> Are the days of prophets, when the use of intelligence was limited, at
> times, to the choice of a "camp," to be regretted? The most simplistic
> divisions—for or against revolution, capitalism, alienation, "symbolic
> force," self-management, and so on—were enough to separate the
> good from the bad without any further examination of the issue being
> necessary. . . . A sinister time, in truth, when the divisions between in-
> tellectuals, true professional ideologues, and experts riveted to their
> administrative careers enabled everyone to avoid the decisive ques-
> tions. (139)

Clearly, we need more than this to tease out the symptomatics of radical
environmentalism and its quite considerable appeal. Here Fredric Jameson's
recent observations on the renascence of ecology and the idea of nature in
postmodern society are particularly suggestive. Jameson argues that ecology in the postmodern moment operates as a genuinely utopian figure for a
longed-for "outside" to global capitalism (to this extent "ecology" remains
tied to the rather different category of "nature" and thus also remains something of a "modern" rather than properly "postmodern" category) *and* at the
same time functions as an index of the failure of postmodern society to
achieve that end. As Jameson puts it in *The Seeds of Time*, in terms with some
utility for exposing what he would call the political unconscious of Ferry's
position, "It seems to be easier for us today to imagine the thoroughgoing
deterioration of the earth and of nature than the breakdown of late capitalism."[12] What has happened, according to Jameson, is a sort of flip-flop of
outside (nature) and inside (the economic, the social) under postmodernism

so that what was formerly "second nature" (the ideologically naturalized economic and social relations of capitalism) has now become the first nature whose end it is impossible to project. Meanwhile, "ecology" has become what Jameson would call an "ideologeme" of postmodern culture, one that trades on the residual, modernist utopian charge of the concept of "nature" while reproducing the systemic logic of the postmodern itself. Nature, Jameson writes, is surely "the strong final term and content of whatever essence or axiomatic . . . whatever limit or fate may be posited." In this sense, the end of nature "is surely the secret dream and longing" of postmodernism understood as the "cultural logic of late capitalism" (46). "Ecology," however, "is another matter entirely," Jameson writes; and what is at question is whether the "nature" of postmodern ecology "is in any way to be thought of as somehow the same as that older 'nature' at whose domestication if not liquidation all Enlightenment and post-Enlightenment thought so diligently worked" (47).

As Jameson notes—and one can't help but recall the expressed or residual misanthropy of some deep ecologists in this connection—very much to the point here is how concepts of nature are always inseparable from those of *human nature*. "A discipline necessarily directed toward the self and its desires and impulses; the learning of new habits of smallness, frugality, modesty, and the like; a kind of respect for otherness that sets a barrier to gratification"—all of these, Jameson reminds us, are "the ethical ideas and figures in terms of which new attitudes toward the individual and the collective self are proposed by (postmodern) ecology" (47). In Jameson's view, then, deep ecology would exemplify "a self-policing attitude," a "new style of restraint and ironic modesty and skepticism about the collective ambitions" of an earlier, modern "Promethean Utopianism" that was of a piece with revolutionary politics, and whose last gasp was the counterculture movement of the 1960s (48). In this light, the thou-shalt-not biocentrism of deep ecology is revealed to be of a piece with a broader "contemporary authoritarianism" in postmodern society, in which a "general pessimism, political apathy, the failure of the welfare state or of the various social democracies—all can be enlisted as causes in a general consent to the necessity for law-and-order regimes everywhere" (48). For Jameson, "such regimes, which it may not be inappropriate to characterize as neo-Confucian . . . finally prove to be based on a renewed conception of human nature as something sinful and aggressive that demands to be held in check for its own good" (49).

Jameson's analysis underscores that there exist both a useful way and a not so useful way to make the point Ferry intends about the overly zealous holism and "antimodernism" of contemporary environmentalism as exemplified by deep ecology. Like Ferry, Jameson is essentially a modernist, if a much more ambivalent and complicated one, and he would defend the modernist Prometheanism necessary for political change in terms more welcome to Ferry than to Hans Jonas. But Jameson's materialism helps us see what is scrupulously avoided in Ferry's defense of liberal humanism: the very direct relations between concepts such as democracy in civil society and the economic structures that materialize or prevent them. In that light, Jameson's analysis helps us tease out not only the "new frugality" of contemporary environmentalism but also the positive content, the material ground and effect, of what Ferry presents as the wholly negative, open-ended, and "free" character of liberal humanism.

. . .

The essential conservatism of Ferry's position is hard to spot at first because his framing of "the new ecological order" sets against the "fundamentalism" and moral Puritanism of contemporary environmentalism the apparent openness and commitment to change—the "*uprootedness,* or *innovation*" as he puts it (xxi)—of the liberal humanist tradition he defends. According to the blend of Rousseau and Kant with which Ferry identifies himself, the "*humanitas*" (of the human) "resides in his freedom, in the fact that he is undefined, that his nature is to have no nature but to possess the capacity to distance himself from any code within which one may seek to imprison him. In other words: his essence is that he has no essence."

"Romantic racialism and historicism are thus inherently impossible," Ferry continues. "For what is racism at its philosophical core if not the attempt to define a category of humans by its essence?" (5). There seems much to admire here and very little to condemn. Unfortunately—as with his concept of democracy—the reality is that Ferry's notion of the human is a good deal less "open" and "innovative" than it first appears. For as we shall see, even though Ferry condemns racism for its attempt to define a category of beings by its essence, this is precisely what his liberal humanist *speciesism* does in relation to nonhuman others in his critique of animal rights philosophy.

One of the fundamental problems with Ferry's discussion of animal rights is foreshadowed by the reliance on simple oppositions of ideal types that we have already seen in his discussion of radical environmentalism; Ferry constantly presents as differences in *kind* what are maintainable only as differences in *degree*. In the case of animal rights, he consistently *over*states the extent to which "the animal is programmed by a code which goes by the name of 'instinct'" (5), and he consistently *under*states the extent to which new work in ethology has shown that many nonhuman animals demonstrate degrees of the volition, free will, and abstraction that Ferry is at great pains to protect as the sole domain of the human. Similarly, he exaggerates in saying that the human being "is *nothing* as determined *by nature*" (9), not bound by instinct, biological needs and intolerances, by sexuality, the body, and so on.

This is not to suggest that Ferry's treatment of animal rights philosophy is as harsh as his attack on deep ecology. Ferry clearly is genuinely concerned with the ethical call on us of nonhuman animals, and he is at some pains to try to do justice to the ethical relevance of the fact that animals (to borrow Heidegger's formulation, which I will scrutinize in the next chapter), if they are not "man" (as he puts it), are also not "stone." Ferry's relative receptivity to animal rights philosophy is less surprising, however, when we remember that the philosophical basis for animal rights as put forward by its two most important practitioners—Peter Singer (in *Animal Liberation*) and Tom Regan (in *The Case for Animal Rights*)—is based squarely in the liberal philosophical tradition of utilitarianism (Singer) and Kantianism (Regan). As Ferry correctly notes, the animal rights argument "*is inscribed in a democratic framework:* in the tradition of Tocqueville, it counts on the progress of 'the equality of conditions,' so that, after the blacks of Africa, animals in turn enter the sphere of rights" (27).

The philosophical basis for animal rights that Singer articulates in *Animal Liberation*—often called the founding text of the animal rights movement—is relatively easy to state and follows Jeremy Bentham's well-known challenge to what would come to be called *speciesism*. In Bentham's words, "What else is it that should trace the insuperable line? Is it the faculty or reason, or perhaps the faculty of discourse? But a full-grown horse or dog is beyond comparison a more rational, as well as a more conversable animal, than an infant of a day or a week or even a month old. But suppose they were otherwise, What would it avail? The question is not, Can they *reason?* nor Can they *talk?* but, Can they *suffer?*" (qtd. in Ferry, 27).

For Singer, beings who have a capacity to suffer—and suffering is very broadly construed here, including not only physical pain but also psychological pain, anticipatory duress, and the like—have a demonstrable *interest* in avoiding suffering; and that means that such beings have a *right* to have those interests protected, to be regarded morally as ends in themselves and not, as Regan puts it (in a phrase with some resonance for specifying his revisionist relationship to Kant's "indirect duty" view), "as mere 'receptacles' of valuable experiences" for humans.[13] And all of this is true, both Singer and Regan argue, *regardless of the species of the being in question.*

If we propose a criterion *other than* suffering for distinguishing between beings who deserve ethical consideration and those who don't—as the humanist tradition has a long history of doing—then the problem, as Singer points out, is that "whatever the test we propose as a means of separating human from non-human animals, it is plain that if all non-human animals are going to fail it, some humans will fail as well."[14] This rejoinder, in turn, is often met by appeals to the *potential* of the human infant, in time, to outstrip her animal counterparts in intelligence, language, and so on, thus achieving a difference not only in degree but *in kind* from the nonhuman animal. But the problem with this retort, of course, is that a significant number of humans—the severely handicapped, say, or the hydrocephalic child—do not possess those capacities and never will. "Why," Singer asks, "do we lock up chimpanzees in appalling primate research centers and use them in experiments that range from the uncomfortable to the agonising and lethal, yet would never think of doing the same to a retarded human being at a much *lower* mental level?" (6). The only answer, Singer argues, is that we are not really using the subject's actual capacities to decide the matter of ethical consideration here, but instead are adjudicating the matter solely based on species. And to do *that* is to indulge in *speciesism*, which—like its cognates racism, sexism, and classism—discriminates against an other based only on a generic description and not on what we actually know about its needs, interests, and capabilities.

This does not mean, as opponents of animal rights often caricature the position, that nonhuman animals have *the same* rights as humans; indeed, both Singer and Regan go to some lengths to rebut this common misunderstanding.[15] Nor does it mean (as Mary Midgley perceptively notes) that all nonhumans necessarily have the same rights. After all, an adult chimpanzee probably has more in common with us than with the trout or chickadee with which it shares the generic classification "animal"; and each of us will have particular behavioral needs and interests that are probably not all that relevant to the others.[16] As Singer points out in a somewhat mischievous example, confining a herd of otherwise well cared for cows to Albany County for a week probably would not infringe on their interests; doing the same to the human inhabitants of Albany County, based on what we know about their physical and psychological needs, probably would. What the animal rights position *does* say, however, is that *all* beings with demonstrable interests (Singer) or "inherent value" (Regan) have a *fundamental* right to avoid suffering that must be respected, regardless of their species. And at that point, the subject of debate usually becomes where to draw the line; cats and elephants and dolphins seem to have clear standing, but do fish? Is the capacity to suffer physical pain enough, or do we need more to grant rights? No matter our answer to that question, as one commentator has pointed out, "in legal or moral discourse we are virtually never able to draw clear lines." But that does not mean (to short-circuit yet another rhetorical strategy of speciesism) "that drawing a line anywhere, arbitrarily, is as good as drawing one anywhere else."[17]

The problem with Singer's animal liberation position, from Ferry's point of view, is that it is based on a faulty fundamental assumption. In an articulation that expresses the very core of his difference with animal rights, deep ecology, and ecofeminism, Ferry writes: "The fundamental difference that separates utilitarianism from the humanism inherited from Rousseau and Kant" is that for the latter "*it is, on the contrary, the ability to separate oneself from interests (freedom) that defines dignity and makes the human being alone a legal subject*" (32). Singer "*never* considers the criteria of freedom defined as the faculty to separate oneself from nature, to resist selfish interests and inclinations" (36). Here again, Ferry is right to alert us to the danger of naturalism in ethics harbored by Singer's thesis. As the ecofeminist Deborah Slicer has pointed out in a somewhat different key, part of the problem with Singer's position is endemic to the liberal justice tradition in moral philosophy of which it is a part. It holds "an 'essentialist' view of the moral worth of both human beings and animals" because it proposes "a single capacity— the possession of interests" (or Singer's "suffering") "for being owed moral consideration," thereby excluding from ethical relevance anything *other* than the specific criterion for the interest in question, whether it is the subject's specific ontogeny, its location or ecological role, its gender, and so on.[18] Slicer's point is well taken, but it is hard to see how *Ferry* could deploy such a critique, since his own basis for maintaining a categorical distinction between the human and nonhuman animal is to posit, precisely, a *single* and defining characteristic ("freedom") as the criterion for ethical consideration.

It is in part to meet the sort of objection to "essentialism" raised by Slicer—and in a different register, by Ferry's objection to Singer's biologism—that Tom Regan, in *The Case for Animal Rights*, critiques the utilitarianism of Singer and broadens the concept of what he calls "inherent value" beyond the emphasis on suffering alone. As Regan puts it,

We are each of us the experiencing subject of a life, a conscious creature having an individual welfare that has importance to us whatever our usefulness to others. We want and prefer things, believe and feel things, recall and expect things. And all these dimensions of our life, including our pleasure and our pain, our enjoyment and suffering, our

satisfaction and frustration, our continued existence or our untimely
death—all make a difference to the quality of our life as lived, as experienced, by us as individuals. As the same is true of those animals
that concern us (the ones that are eaten and trapped, for example),
they too must be viewed as the experiencing subjects of a life, with inherent value of their own.[19]

But Regan's position, though it moves us beyond the biologism Ferry criticizes, does not exactly dispose of the problem of essentialism that Slicer
finds not only in Singer and Regan but in Ferry as well. Indeed, in this light
the problem with animal rights philosophy is not that it is antihumanist, but
rather that it is *too* humanist. As Slicer argues, rights theories "reduce individuals to that atomistic bundle of interests that the justice tradition recognizes as the basis for moral considerableness. In effect, animals are represented as beings with the *kind of capacity* that human beings most fully
possess and deem valuable for living a full *human* life" (111). Stephen Zak
captures the problem particularly well:

> Lives don't have to be the same to be worthy of equal respect. One's
> perception that another life has value comes as much from an appre-
> ciation of its uniqueness as from the recognition that it has character-
> istics that are shared by one's own life. (Who would compare the life
> of a whale to that of a marginal human being?) . . . The orangutan can-
> not be redescribed as the octopus minus, or plus, this or that mental
> characteristic: conceptually, nothing could be added to or taken from
> the octopus that would make it the equivalent of the oriole. Likewise,
> animals are not simply rudimentary human beings, God's false steps,
> made before He finally got it right with us. (70)

Zak lucidly locates a fundamental problem with animal rights philosophy in its current state of the art—and it is the problem that links it to Ferry's
supposedly opposite humanism. And even as a critique of animal rights *internal* to the liberal philosophical tradition, Ferry's Kantian-Rousseauian
position encounters all sorts of difficulties. First, his reliance on "freedom"
to serve as the ethical wedge between the human and nonhuman animal
does nothing, despite his gestures to the contrary (42), to address the problem of the "lowest common denominator" raised by animal rights philosophy. Following Ferry, we would be forced to say that the hydrocephalic infant had no interests and rights and could therefore be exploited as pure
means (just as laboratory animals are) because it neither embodies nor has
the capacity for the liberal "freedom" that ensures ethical consideration.

Ferry attempts at the end of his second chapter to forestall this pursuit of his humanism to its logical conclusions, but he can do so only at the price of an utterly question-begging resort to speciesism. "Why sacrifice a healthy chimpanzee over a human reduced to a vegetable state?" Ferry asks. "If one were to adopt the criteria that says there is continuity between men and animals, Singer might be right to consider as 'speciesist' the priority accorded human vegetables. If on the other hand we adopt the criteria of freedom, it is not unreasonable to admit that we must respect humankind, even in those who no longer manifest anything but its residual signs" (42). But of course it *is* "unreasonable," because in this instance Ferry isn't relying on the quality of "freedom" *at all* to ethically adjudicate the matter (the human vegetable, by Ferry's own admission, does not possess this quality) but is using only membership in a given *species*. And this is no better than the racism that is supposedly impossible under Ferry's humanism.

It should come as no surprise, then, that Ferry is unable to satisfactorily address an important issue raised by animal rights philosophy, one I have already touched on: that the discourse and practice of speciesism in the name of liberal humanism have historically been turned on other *humans* as well. (Here Singer would be quick to point out that the first chapter of *Animal Liberation* is titled "All Animals Are Equal, or Why Supporters of Liberation for Blacks and Women Should Support Animal Liberation Too" [1].) To his credit, Ferry seems to recognize this problem. "This distinction between humanity and the animal kingdom seems to carry horrifying consequences in its wake" (12), he writes. "It is impossible to avoid racism and its political consequences if one subscribes to the belief that primitive man cannot attain authentic humanity due to his essence or nature" (13). "But this was not," he continues, "the *Aufklärer's* response" (13); for liberal humanism, "*this difference is not inscribed in a definition, in a racial essence.* We are forced to agree with Musil that a 'cannibal taken from the cradle to a European setting will no doubt become a good European and that the delicate Rainer Maria Rilke would have become a good cannibal had destiny, to our great loss, cast him at a tender age among the sailors of the South Seas'" (14).

But such an example demonstrates not so much, as Ferry thinks, the progressivism of Enlightenment humanism as the question-begging concept of "freedom" in his own critique. For how can Ferry locate the basis of ethical consideration in freedom, defined "by perfectibility, by the capacity to break away from natural or historical determinations" (15), and at the same time praise the way Enlightenment culture recognizes (as in Musil's example) the force of historical determination to wholly shape one's character?[20] It is a case of out of the frying pan and into the fire, for as I have already noted, you don't *need* the argument from "racial essence" to justify oppression if you can control the discourses and institutions that reduce human beings to the status of objects. One's belief that women have more free will and control over the finality of their actions than a nonhuman animal does not prevent the use of the discourse of speciesism in the oppression of women. As with Ferry's deployment of the term "democracy," "freedom" in his humanist lexicon turns out to be a good deal less free—and a good deal more historically and socially specific—than he would have us believe.

This should surprise no one, however, for Ferry's is essentially a "contractarian" model of ethics. As Regan points out, the contractarian view stipulates that those who can understand and freely enter into the ethical contract are protected by it and can also seek protection, under certain conditions, for beings (such as infants, who are unable to understand or sign) whom they care about but who are not themselves signatories. We have, as Kant argues, an "indirect duty" to those other beings—duties *involving* them, but no duties *to* them. In Ferry's formulation, "nature is not an *agent*, a being able to act with the *reciprocity* one would expect of an *alter ego*. *Law is always for men*, and it is for men that trees or whales can become *objects* of a form of respect tied to legislation—not the reverse" (139). "Animals," Ferry argues, "have no rights (as zoophiles would have it), but on the other hand we do have certain, indirect duties toward them, or at least 'on their behalf,'" because the animal "is (or should be) the object of a *certain* respect, a respect which, by way of animals, we *also* pay ourselves" (53–54). In short, from the "indirect duty" point of view, the problem with cruelty toward animals is not that it is a violation of their basic interest in avoiding suffering and their basic right not to be treated as objects; it is rather that "the most serious consequence of the cruelty and bad treatment inflicted on them *is that man degrades himself and loses his humanity*," that cruelty "*can affront or corrupt man's sensibility*" (25). It is "a matter of *politeness* and *civility*" (56).

From Regan's point of view, "contractarianism could be a hard view to refute if it were an adequate theoretical approach to the moral status of human beings" (17). But it isn't, because contractarianism "is very well and good for the signatories but not so good for anyone who is not asked to sign." Nor do those whom the contract presents as equal partners always enter the contract equally "freely." There is nothing in contractarianism to guarantee "that everyone will have a chance to participate equally in framing the rules of morality," and there is nothing in principle to obviate Regan's observation that "might, according to this theory, does make right." "Such a theory takes one's moral breath away," Regan continues, "as if, for example, there would be nothing wrong with apartheid in South Africa if few white South Africans were upset by it" (17).

But how does a society decide who gets to be covered in such a contract and who doesn't? And how do we escape what looks like an inescapable dilemma posed by Ferry's "might makes right" contractarianism on the one side and what he would see as Singer's and Regan's naturalism and biologism on the other? In other words, if we want to avoid the political perils of answering the question, "Who gets ethical treatment?" with "whoever fits the particular prejudices and bigotries of a particular society,"[21] then how do we *also* avoid what seems like the only viable alternative: to locate a noncontingent natural ground (Singer's "suffering," say) that is outside—and not properly subject to—the contingency of the social contract?

The way out of this dilemma resides, I think, in adopting a thoroughly pragmatist approach—but a pragmatism, as I have argued in some detail elsewhere, renovated by sustained engagement with and enrichment by postmodern theory.[22] Such an approach (in line with the pragmatist tradition generally) declares out of bounds any *representationalist* account of how we might "ground" the ethical standing of being X in some more empirically "true" understanding of its actual nature. To reject representationalism is to reject "the idea," as Richard Rorty puts it, "that inquiry is a matter of finding out the nature of something which lies outside the web of beliefs and desires," in which "the object of inquiry—what lies outside the organism—has a context of its own, a context which is privileged by virtue of being the object's rather than the inquirer's" (*Objectivity*, 4, 96). In this way pragmatism "switches attention from the 'demands of the object' to the demands of the purpose which a particular inquiry is supposed to serve," with the effect that "now one is debating what purposes are worth bothering to fulfill, which are more worthwhile than others, rather than which purposes the nature of humanity or of reality obliges us to have. For antiessentialists, all possible purposes compete with one another on equal terms, since none are more 'essentially human' than others" (*Objectivity*, 110).

In this light, the problem with Ferry's antinaturalism is that it is not *anti* enough. That is, Ferry does not rigorously abide by his own demand that we stop basing ethics in some more or less transparent understanding of the "natural" qualities of a being and understand that ethics is about *concepts* and not about *objects*—that it is, to use Jean-François Lyotard's terminology, a language game. The liberal democratic tradition has, as Ferry proudly claims, adjudicated the matter of ethical consideration based not on naturalism, or biology, or race, or any other "given" characteristic, but rather on the identifiable qualities and characteristics of ethical subjects irrespective of what John Rawles calls their "accidental" qualities. Most notable among these for Ferry, of course, is "freedom." But just what this "freedom" consists of is the subject of considerable slippage and vexation in his argument. In some places he holds that it is the ability "*to separate oneself from interests*" (32), "the faculty to separate oneself from nature, to resist selfish interests and inclinations" (36).

But as I noted in the opening pages, the problem with this formulation is that it readily applies to several nonhuman animals as well. As any number of very prominent studies in field ecology, cognitive ethology, and linguistic production in great apes over the past twenty years have shown, the "defining" characteristics of the distinctly human—language, tool use, tool *making*, social behavior, altruism, and so on—have been found to be not so defining after all. Whether in academic studies such as Marc Bekoff and Dale Jamieson's *Interpretation and Explanation in the Study of Animal Behavior*, Marian Stamp Dawkins's *Through Our Eyes Only? The Search for Animal Consciousness*, and Donald R. Griffin's *Animal Minds* or in rather more popular texts like Paola Cavalieri and Peter Singer's *The Great Ape Project*,[23] it has become clear that some nonhuman animals—chimpanzees, for instance— "share with us tool-making and tool-using capacities, the faculty for (nonverbal) language, a hatred of boredom, an intelligent curiosity towards their environment, love for their children, intense fear of attack, deep friendships, a horror of dismemberment, a repertoire of emotions and even the same capacity for exploitive violence that we so often show towards them."[24] This list can be—and has been—expanded in great detail to include self-awareness, the ability to regularly engage in both deceptive and altruistic behavior, and many another quality thought for centuries to be exclusively human. This widely disseminated body of work makes it abundantly clear that many nonhuman animals "separate themselves from selfish interests and inclinations" all the time, as a matter of course—that "freedom," in fact, is key to the evolutionary success of many nonhuman species.

In other places Ferry seems to realize the futility of the "freedom from instinct" strategy and offers instead a tortured argument that "because of this capacity to act in a nonmechanical fashion, oriented by a goal" the animal is an "*analogon* of a free being" (46) and that "life, defined as 'the faculty to act according to the representation of a goal,' is an *analogue of freedom*" (54). The problem with *this* rather desperate gambit, however, is that it assimilates "life" in general to goal-oriented behavior while declaring beside the point the complex forms of social interaction, communication, and self-awareness that seem very much to the ethical point. Such a formulation would force us to crudely lump the mountain gorilla with the amoeba (both are instances of "goal-oriented" "life"). But clearly the mountain gorilla—for reasons not legible by this formulation—has much more in common with *Homo sapiens*.

Finally, Ferry attempts to raise the bar of "freedom" and the "distinctly human" one last time. In almost the only place in the book where he seems vaguely aware of the explosion of revisionist work in ethology in the past twenty years, he writes:

> One can cite the suicide of whales—an indication that they too can distance themselves from their natural tendency—the language of monkeys and dolphins, the capacity of certain animals to manipulate tools in order to realize their objectives, not to mention canine devotion or feline independence. . . . The problem, of course, is that this separation from the commandments of nature is not transmitted from *one generation to the next* as a history. A separation from natural norms only becomes evident when it engenders a cultural universe. (6)

In his desperate attempt to maintain the species barrier, Ferry first tries out "freedom from instinct," then freedom versus the "analogue" of "goal-oriented behavior," and finally cultural transmission. Aside from begging the question of *why* the transmission of cultural behavior from one generation to the next is ethically *fundamental* ("freedom from nature," the crux of the whole Kantian-Rousseauian position Ferry espouses, would fall well to this side of "cultural transmission"); and aside from providing a disturbing echo of similar statements made during the past two centuries about "primitive" societies, for whom culture seems to be a form of constraint and continuity, and which would thereby run afoul of Ferry's ethnocentric view of cultural as "innovation" and "uprootedness"; and aside from leaving completely untouched the ethical relevance of the "lowest common denominator" problem raised by animal rights—aside from all that, Ferry seems to have his facts wrong. For as Jane Goodall points out,

> We can speak of the history of a chimpanzee community, where major events—an epidemic, a kind of primitive "war," a "baby boom"—have marked the reigns of the five top-ranking alpha males we have known. And we find that individual chimpanzees can make a difference to the course of chimpanzee history, as is the case with humans. . . .
> Chimpanzees, like humans, can learn by observation and imitation, which means that if a new adaptive pattern is "invented" by a particular individual, it can be passed on to the next generation. Thus we find

that while the various chimpanzee groups that have been studied in different parts of Africa have many behaviors in common, they also have their own distinctive traditions. This is particularly well-documented with respect to tool-using and tool-making behaviours. Chimpanzees use more objects as tools for a greater variety of purposes than any creature except ourselves, and each population has its own tool-using cultures.[25]

One can only imagine that Ferry's response to *this* would be to raise the bar once again, so that only those who have read all of Kant's *Critiques* and passed the exegesis on to their grandchildren would be eligible for ethical consideration!

More promising here, I think, is the pragmatist approach to the problem I invoked earlier, within which it is perfectly possible to argue that taking account of the ethical relevance of the work of ethologists like Goodall—or of the biologists Humberto Maturana and Francisco Varela in the next chapter—*does not mean committing ourselves to naturalism in ethics*. From a pragmatist point of view, all it means is that, in the historically and socially contingent discourse called "ethics," we are obliged—precisely *because* ethics cannot ground itself in a representationalist relation to the object—to apply consistently the rules and norms we devise for determining ethically relevant traits and behaviors, without prejudice toward species or anything else. The strength of this position lies (as Stanley Cavell or Gianni Vattimo might say) precisely in its "weakness." I will address this epistemological issue in much greater detail in the next chapter, but for now it is enough to note that we need not cling to any empiricist notion about what Goodall or anyone else has discovered about nonhuman animals—any more than we need to do the same for our knowledge of human beings—to insist that when our generally agreed-on markers for ethical consideration are observed in species other than *Homo sapiens*, we are obliged to take them into account equally and to respect them accordingly. This amounts to nothing more than taking the humanist conceptualization of the problem at its word and being rigorous about it—and then showing how humanism must, if rigorously pursued, generate its own deconstruction once these "defining" characteristics are found beyond the species barrier. But this, of course, is precisely what Ferry is unable and unwilling to do. This is not to say that the ethical question of the animal ends here—and indeed, the next chapter will explore dramatically different alternatives for examining this question with greater resonance and subtlety. Rather, it means that the self-deconstruction of humanism is only where that question begins to open onto the various registers of its full complexity.

In the end, then, it is Ferry's "human," and not, as he argues, the nonhuman animal, who is "the enigmatic being," the "dreamed object"—"enigmatic" because incoherent, and "dreamed" because an imaginary subject, a fantasy. And yet, for all that, quite familiar. For as we have already seen, the humanist concept of subjectivity is inseparable from the discourse and *institution* of speciesism, which relies on the tacit acceptance—and nowhere more clearly (as Slavoj Žižek has noted) than in Ferry's beloved Kant—that the full transcendence of the "human" requires the sacrifice of the "animal" and the animalistic, which in turn makes possible a symbolic economy in which we can engage in a "noncriminal putting to death" (as Derrida puts it) not only of animals, but other *humans* as well by marking *them* as animal.[26] It may be, as Ferry argues, that ethics is always ineluctably human, always about human concepts and not about objects; but what Ferry's concept of "the human" fails to do acknowledge—indeed, his project depends on its disavowal—is how this constitutive and finally desperate repression only reveals all the more surely what Žižek calls "humanism's self-destructive dimension"—one that I will examine in painstaking detail later in discussing *The Silence of the Lambs*.[27] As Žižek puts it, "The subject 'is' only insofar as the Thing (the Kantian Thing in itself as well as the Freudian impossible-incestuous object, *das Ding*) is sacrificed, 'primordially repressed.' . . . This 'primordial repression' introduces a fundamental imbalance in the universe: the symbolically structured universe we live in is organized around a void, an impossibility (the inaccessibility of the Thing in itself)."[28] "Therein," Žižek continues, "consists the ambiguity of the Enlightenment"; the transcendence of the Enlightenment subject is shadowed by "a fundamental prohibition to probe too deeply into the obscure origins, which betrays a fear that by doing so, one might uncover something monstrous" (136).

We could scarcely do better than Žižek's characterization to provide a thumbnail psychoanalysis of Ferry's *New Ecological Order*. But when we remember, with Derrida, that the effectiveness of the discourse of species, when applied to social others of *whatever* sort, relies on first taking for granted the *institution* of speciesism—that is, on the ethical acceptability of the systematic, institutionalized killing of nonhuman others—then it is clear that the ethical priority of the question of the animal, while it may begin with "man" and his self-destructive humanism, does not end there.

CHAPTER 2

In The Shadow of Wittgenstein's Lion

Language, Ethics, and the Question of the Animal

FORMS OF LANGUAGE, FORMS OF LIFE: WITTGENSTEIN, CAVELLI, AND HEARNE

In 1958, toward the end of his *Philosophical Investigations*, Ludwig Wittgenstein set down a one-sentence observation that might very well serve as an epigraph to the debates that have taken place over the past century on animals, language, and subjectivity. "If a lion could talk," Wittgenstein wrote, "we could not understand him."[1] This beguiling statement has often been misunderstood—I'm not sure I understand it myself—and it is only complicated by Wittgenstein's contention elsewhere that "to imagine a language is to imagine a form of life."[2] What can it mean to imagine a language we cannot understand, spoken by a being who cannot speak—*especially* in light of his reminder that "the kind of certainty is the kind of language-game" (*Wittgenstein Reader*, 213)? And earlier still: "If I were to talk to myself out loud in a language not understood by those present my thoughts would be hidden from them" (211). "It is, however, important as regards this observation that one human being can be a complete enigma to another. We learn this when we come into a strange country with entirely strange traditions; and, what is more, even given a mastery of the country's language. We do not *understand* the people. (And not because of not knowing what they are saying to themselves.)" (212).

It is the caginess, if you will, of the muteness of Wittgenstein's lion that rightly catches the attention of Vicki Hearne in her *Animal Happiness*. Hearne—a poet, a renowned horse and dog trainer, and a serious student of the philosophical lineage that runs from Wittgenstein through Stanley Cavell—calls Wittgenstein's statement "the most interesting mistake about animals that I have ever come across," because "lions do talk to some people"—namely lion trainers—"and are understood" (a claim about language that we will have occasion to revisit).[3] What interests her is how Wittgenstein's statement seems—but only seems—to body forth an all too familiar contrast between the confidently transparent intersubjective human community, on the one hand, and the mute, benighted beast on the other. It is this contrast, and this humanism, however, that Wittgenstein is out to trouble, for as Hearne notes, "The lovely thing about Wittgenstein's lion is that Wittgenstein does not leap to say that his lion is languageless, only that he is not talking"—a remark that is "a profundity rarely achieved, because of all it leaves room for" (169). "The reticence of this lion," she continues, "is not the reticence of absence, absence of consciousness, say, or knowledge, but rather of tremendous presence," of "all consciousness that is beyond ours" (170).

What Hearne puts her finger on here—what she finds attractive in the style or posture of Wittgenstein's "mistake"—is the importance of how we face, face up to, the fact of a "consciousness beyond ours." More specifically, what value do we attach to the contention that animals "do not talk, that no bit of their consciousness is informed by the bustle and mediations of the written, the symbolic" (171)? For Hearne, what makes Wittgenstein's intervention valuable is that this darkness or muteness of the animal other is shown to be more a problem for *us* than for the animal. "The human mind is nervous without its writing, feels emptiness without writing," she reminds us. "So when we imagine the inner or outer life of a creature without that bustle, we imagine what we would be like without it—that is, we imagine ourselves emptied of understanding" (171). Thus Wittgenstein's lion "in his restraint remains there to remind us that knowledge . . . comes sometimes to an abrupt end, not vaguely 'somewhere,' like explanations, but immediately"—a fact dramatized for Hearne when the understanding between lion and lion trainer goes wrong. Wittgenstein's lion, "regarded with proper respect and awe, gives us unmediated knowledge of our ignorance" (173).

"Not 'somewhere,' like explanations," is anything but a throwaway phrase in this instance, for it takes us to the very heart of Wittgenstein's transvaluation of philosophical skepticism, one best elaborated by Stanley Cavell. For Cavell, our tendency to see the reticence of Wittgenstein's lion as a lack of subjectivity is symptomatic of nothing so much as "our skeptical terror about the independent existence of other minds"—a terror that is, in a certain sense, about our failure to be god, to be "No One in Particular with a View from Nowhere," as Hearne puts it (*Adam's Task*, 233, 229). And this terror, in turn, drives the fantasy that, through philosophy, we somehow might be. As Hearne writes of "thinkers who like to say that a cat cannot be said to be 'really' playing with a ball because a cat does not seem to know our grammar of what 'playing with' and 'ball' are" (a position, incidentally, that is sometimes attributed to Wittgenstein):

This more or less positivist position requires a fundamental assumption that "meaning" is a homogeneous, quantifiable thing, and that
the universe is dualistic in that there are only two states of meaning in
it—significant and insignificant, and further that "significant" means
only "significant to me." . . . Such positivism of meaning looks often
enough like an injunction against the pathetic fallacy, but seems to me
to be quite the opposite. (*Adam's Task*, 238)

In Hearne's and Cavell's readings, skeptical terror generates certain
philosophical concepts of language and its relation to consciousness and
subjectivity that it is Wittgenstein's business to subvert—and subvert in a
rather peculiar way. As Cavell puts it, what prevents our understanding of
animals—take Wittgenstein's lion as only the most hyperbolic example—
"is not too much skepticism but too little" (qtd. in Hearne, *Adam's Task*,
114). For Cavell, the philosophical false start that Wittgenstein wants to
reroute is "the (skeptic's) idea that the problem of the other is the problem
of *knowing* the other," when in fact one of the most valuable things about our
encounter with the supposedly "mute" animal is that it "sooner makes us
wonder what *we* conceive knowledge to be" (qtd. in *Adam's Task*, 114; my
emphases). If we follow Wittgenstein's lead, Cavell argues, "One is not en-
couraged . . . to go on searching for a something—if not a mechanism, or
an image, then a meaning, a signified, an interpretant—that explains how
calls reach what they call, how the connection is made," but rather "to determine what keeps such a search going (without, as it were, moving).
Wittgenstein's answer, as I read it, has something to do with what I understand as skepticism, and what I might call skeptical attempts to defeat skepticism." For Cavell, Wittgenstein not only "shows us that we maintain un-
satisfiable pictures of how things must happen," he also forces us to think
through "why we are, who we are that we are, possessed of this picture."[4]

Wittgenstein's specific intervention, then—his "skeptical attempt to defeat skepticism"—is to turn philosophical skepticism back on itself, back on
the human. Hence the project of what is often remarked as Wittgenstein's
conventionalism is in no small part "to make us dissatisfied with the idea of
universals as explanations of language."[5] Philosophy may always seem to
want to situate itself outside the noise and contingency of language games,
"but it depends on the same fact of language as do the other lives within it":
that "it cannot dictate what is said *now*, can no more assure the sense of what
is said, its depth, its helpfulness, its accuracy, its wit, than it can insure its
truth to the world" (*Claim*, 189). As Hearne puts it in an essay on the famous
language experiments with Washoe the chimpanzee, "The issue of what

Washoe is doing, what condition of language we are dealing with, is not an intellectual problem, a puzzle." If Washoe uses language and remains dangerous despite that (which she most certainly does), "then I may be thrown into confusion . . . and may want to deny Washoe's personhood and her language rather than acknowledge the limits of language—which can look like a terrifying procedure" (*Adam's Task*, 39).

This means, in Cavell's words, that "we begin to feel, or ought to, terrified that maybe language (and understanding, and knowledge) rests upon very shaky foundations—a thin net over an abyss" (*Claim*, 178). And it is also an apt description of what Wittgenstein has in mind when he says, famously, that to imagine a language is to imagine a "form of life." As Hearne puts it, "One can hang out with people who speak no English and learn something of which objects are meant by which words. What is much harder to know, what you have to be deeply, genuinely bilingual to know, is what the object or posture itself means. I may know that *shlumah-ney* means what I call 'candle,' but not whether candles are sacred to my 'informants,' and not such things as whether to ask permission to use the candle to read in bed at night" (*Happiness*, 170). For Cavell, "It is such shades of sense, intimations of meaning, which allow certain kinds of subtlety or delicacy of communication: the communication is intimate, but fragile. Persons who cannot use words, or gestures, in these ways with you may yet be in your world, but perhaps not of your flesh" (*Claim*, 189).

At this point in the argument, the Wittgensteinian lineage seems promising indeed for our ability to reconjugate the relations between language, species, and the question of the subject, not least because Wittgenstein's conventionalism appears to more or less permanently unsettle the ontological difference between human and animal, a difference expressed in the philosophical tradition by the capacity for language: first, by holding that that ontological difference is itself constituted by a language that cannot ground and master a world of contingency via "universals," and second, by showing that language does not answer the question, What's the difference between human and animal? Rather, it keeps that question alive and open by insisting that the differences between participants in specific language games and those "not of their flesh" may be as profound as those usually taken to obtain between the human *as such* and the animal *as such*—as if there were, any longer, any such thing *as such*.

What Wittgenstein's account makes possible, in other words, is what we might call a conventionalist understanding of the shared dynamics of a world building that need not, in principle, be tied to species distinctions *at all*. On this account, not *the* world but simply *a* world emerges from building a shared form of life through participation in a language game. And indeed this is the direction in which Hearne has taken Wittgenstein's cue in her writings on how the shared language of animal training makes possible a common world between beings with vastly different phenomenologies. For Hearne, "training creates the kind of knowledge all talking does, or ought to do—knowledge of the loop of intention and openness that talk is, knowledge of and in language" (*Adam's Task*, 85). And if "the sketchiness of the tokens of this language game" might look to a scientist like "the wildest sort of anthropomorphizing"—as when a trainer says a certain dog has a mischievous sense of humor—what has to be remembered is that "a reason for trying to get a feel for a dog-human language game is that it sharpens one's awareness of *the sketchiness of the tokens of English*" (*Adam's Task*, 71–72; emphasis mine). "With horses as with dogs," she continues, "the handler must learn to believe, to 'read' a language s/he hasn't sufficient neurological apparatus to test or judge, because the handler must become comprehensible to the horse, and to be understood is to be open to understanding, much more than it is to have shared mental phenomena. It is as odd as Wittgenstein suggested it is to suppose that intersubjectivity depends on shared mental phenomena" (*Adam's Task*, 106). What it depends on instead is the "flow of intention, meaning, believing," the "varied flexions of looped thoughts," which is why "the behaviorist's dog will not only seem stupid, she will be stupid. If we follow Wittgenstein in assuming the importance of assessing the public nature of language, then we don't need to lock a baby up and feed it by machine in order to discover that conceptualization is pretty much a function of relationships and acknowledgement, a public affair" (*Adam's Task*, 58).

And yet, in both Hearne and Cavell, what I will characterize much too quickly here as a kind of humanism, a palpable nostalgia for the human, returns through the back door to severely circumscribe the ethical force of the shared world building with animals that at first seems to be promised by their appropriation of Wittgenstein, leaving the animal ethically if not phenomenologically benighted and the human insufficiently interrogated by the encounter. The clunkiest symptom of this, perhaps, is the social contract theory of rights that Hearne borrows, at least in part, from Cavell (who in turn borrows it largely from John Rawls)—a theory we have already found wanting in the previous chapter.[6] To put it very schematically, the contractarian view holds that

> morality consists of a set of rules that individuals voluntarily agree to
> abide by, as we do when we sign a contract. . . . Those who understand

and accept the terms of the contract are covered directly; they have rights created and recognized by, and protected in, the contract. And these contractors can also have protection spelled out for others who, though they lack the ability to understand morality and so cannot sign the contract themselves, are loved or cherished by those who can. . . . As for animals, since they cannot understand the contracts, they obviously cannot sign; and since they cannot sign, they have no rights. [B]ut those animals that enough people care about (companion animals, whales, baby seals, the American bald eagle), though they lack rights themselves, will be protected because of the sentimental interests of people. I have, then, according to contractarianism, no duty directly to your dog or any other animal, not even the duty not to cause them pain or suffering; my duty not to hurt them is a duty I have to those people who care about what happens to them.[7]

This is the view, derived from Kant, that is expounded by Hearne, nearly to the letter, in an essay originally published under the title "What's Wrong with Animal Rights?" In order to be in a rights relation with another, she argues, "the following minimum conditions must hold": "I must know the person," "the person must know me," "the grammar of the reciprocal possessive must apply," and "both of us must have the ability to conceive the *right* in question itself" (*Happiness*, 209). For Hearne, "If I do not own you, own up to you, then I do not acknowledge you, I repudiate you. You cannot have interests or rights in relationship to me unless we own each other" (*Happiness*, 206).

Not surprisingly, this leads Hearne into all sorts of tortured formulations in which she seems to forget everything she has spent the better part of her career teaching us about nonhuman others and the worlds we may inhabit with them: "The kind of possession I have in mind is not like slavery. It does not bind one party while freeing the other. . . . [I]f I abuse my dog on the grounds that she is my dog, then I do not, at the moment at least, in fact own the dog, am not owning up to what goes into owning a dog, do not understand my own words when I say I own the dog and can therefore do as I please with her" (208). Or again, writing of her famous Airedale, "Drummer can speak to his owner, but he cannot speak either to or of the state. Therefore the state cannot grant rights to Drummer, cannot be *his* state. Hence it is not an incidental or accidental but a central fact that in practice the only way a dog's rights are protected, against neighbors or the state, is *by way of an appeal to the owner's property rights in the dog*" (212). Of course, this is tantamount to simply wishing that all owners will be "good" ones. And if they are not—if an owner decides to set his dog on fire, instead of a chair or table, its equivalent under the law (as property)—then doesn't this beg the question that the whole *point* of granting rights to the animal would be to *directly* recognize and protect it (as we do with the guardianship of the child) against such an owner who decides to forget or abrogate, for whatever reason, what "ownership means"?

In addition to the usual objections associated with the contractarian view of ethics, which I will list briefly in a moment, matters are not helped any in *Cavell's* case by his (admittedly) iconoclastic reading of Wittgenstein's concept of "forms of life." In contrast to what he calls the dominant "ethno-logical" or "horizontal" reading of this moment in Wittgenstein, Cavell emphasizes the "biological or vertical sense," which "recalls differences be-tween the human and so-called 'lower' or 'higher' forms of life, between, say, poking at your food, perhaps with a fork, and pawing at it, or pecking at it." Here—and I will return to this figure below in discussing Jacques Der-rida's reading of Heidegger—"the romance of the hand and its apposable thumb comes into play, and of the upright posture and of the eyes set for heaven; but also the specific strength and scale of the human body and of the human senses and of the human voice."[8] Cavell takes issue with those who see Wittgenstein's conventionalism as an automatic refutation of skepti-cism, a reading in which "the very existence of, say, the sacrament of mar-riage, or of the history of private property, or of the ceremony of shaking hands, or I guess ultimately the existence of language, constitutes proof of the existence of others" (*This New*, 42)—a position that would be consonant with the "hard" conventionalist reading of a Richard Rorty or a Stanley Fish. Cavell's emphasis not on "*forms* of life," but forms of *life*" intends in-stead to "mark the limit and give the conditions of the use of criteria as ap-plied to others" (*This New*, 42–43), with the larger aim of contesting the "sense of political or social conservatism" that for many readers attends Wittgenstein's *Philosophical Investigations* (*This New*, 44). The idea here, from Cavell's vantage, is that by positing a figure of the human form of *life* not reducible to the immanence ("forms") of language games, Wittgenstein provides a yardstick, or at least a background, against which those language games (private property, for instance) may be judged as desirable or want-ing.[9] What Cavell calls "the practice of the ordinary"—being responsible to the everyday details of a specific "form of life"—"may be thought of as the overcoming of iteration or replication or imitation by repetition, of count-ing by recounting, of calling by recalling. It is the familiar invaded by an-other familiar" (*This New*, 47).

And yet the problem is that this moment—and it is for Cavell the

moment of ethics—is accompanied by a strong return to the very humanism that his phenomenological speculations had promised to move us beyond. If we take seriously the ethnological or conventionalist sense of
Wittgenstein's "forms of life," as Cavell realizes we must, then we are faced
very quickly with this ethical dilemma: the balkanization of language games
promises to circumscribe ever more tightly those who share my world—
those who are, to use Cavell's phrase, "of my flesh." The verticality of language games that Wittgenstein insists on strengthens the shared ethical call
of those *within* the game, but only at the expense of weakening the ethical
call in relation to those who speak in other tongues (hence Cavell's worries
about Wittgenstein's conventionalist conservatism).

It is as if to arrest this runaway mitosis of the linguistic and ethical field
that both Hearne and Cavell reintroduce a certain figure of the human familiar to us from the liberal tradition. In Hearne, for example, the language
of animal training provides a shared language game, and hence a shared
world, between trainer and animal; but ethically speaking, that symmetry of
relation, as she describes it, is belied by the radical *asymmetry* that obtains
when the ethical relation of rights is properly expressed, as she argues, in the
institution of property ownership. And it is not at all clear, of course, that
we have any ethical duty whatever to those animals with whom we have not
articulated a shared form of life through training or other means. Hearne's
contractarian notion of rights only reinforces the asymmetrical privilege of
the ethnocentric "we," whereas the whole point of rights seems to be that it
affords protection of the other *exactly in recognition of* the dangers of an
ethnocentric self-privileging that seems to have forgotten the fragility and
"sketchiness" of its *own* concepts, its *own* forms of life, in the confidence with
which it restricts the sphere of ethical consideration.

In Cavell, things play out rather differently, specifically in his rendering
of the human "form of life" over against "the so-called 'lower'" forms. In
The Claim of Reason, the slippage from human to human*ist* and the ethical
foreclosure that attends it is especially pronounced. Investigating the biological or "vertical" sense of "forms of life" as "the background against
which our criteria do their work; even, make sense," Cavell quotes Wittgenstein: "Only of a living human being and what resembles (behaves like) a living human being can one say: it has sensations; it sees; is blind; hears; is deaf;
is conscious or unconscious" (83). Cavell takes this and other similar moments in Wittgenstein to mean that it is not any conventionalist criterion
but our biological form of life that leads us to such attributions, so that "to
withhold, or hedge, our concepts of psychological states from a given creature"—exactly the position taken by Thomas Nagel in his well-known essay

"What Is It Like to Be a Bat?"—"is specifically to withhold *the source of my idea* that living beings are things that feel; it is to withhold myself, to reject my response to anything as a living being; to blank so much as my idea of anything as *having a body*" (83; first emphasis mine). When we do so,

> There is nothing to read from that body, nothing the body is *of;* it does not go beyond itself, it expresses nothing. . . . It does not matter to me now whether there turn out to be wheels and springs inside, or stuffing, or some subtler or messier mechanism. . . . What this "body" lacks is *privacy.* . . . Only *I* could reach that privacy, by accepting it as a home of my concepts of the human soul. When I withdraw that acceptance, the criteria are dead. . . . And what happens to me when I withhold my acceptance of privacy—anyway, of otherness—as the home of my concepts of the human soul and find my criteria to be dead, mere words, word-shells: I said a while ago in passing that I withhold myself. What I withhold myself from is my attunement with others—with all others, not merely with the one I was to know. (*Claim,* 84–85)

Now there are many things that could be said about this fascinating passage. One might, for example, ask why the sentences on "wheels and springs" do not beg the question that is often raised so forcefully in science fiction—in the film *Blade Runner,* say—about why there *should* be any necessary relation between the phenomenological and ethical issues that attend what we usually denote by the term "human" and the particular physical mechanism of its realization. Or one might point to how phrases such as "nothing the body is *of*" reintroduce the danger of what Daniel Dennett has called the "Cartesian theatre" of a mind (or ego, *cogito,* or, here, "soul") that threatens to evaporate into "no one in particular with a view from nowhere."[10] Or one might argue, as I did in the previous chapter, that a passage like this makes clear why the supposed "weakness" of philosophical conventionalism (its "pragmatism") is precisely its strength. That is, instead of openness to the other depending on a representationalist adequation between otherwise "dead" criteria and the genus of being whose "true" nature allows us to say that those criteria are being properly deployed—in which case we are forced to ask, How much "of our flesh" is flesh enough?—relevant criteria should instead apply consistently and dispassionately across the board, pragmatically, not because certain entities are a priori certain *types of beings.* In this light, the problem is that there is in the foregoing passage no way to stop the difference between "wheels and springs" and "nerves and muscles" from readily rescripting itself not only as the difference between human and android (to stay with the *Blade Runner* example) but also, for our purposes here, as the difference between human and animal.

My larger point, however, is that this "living being" turns out to be a fairly familiar sort of creature after all (as is suggested most pointedly, perhaps, by the discourse of "privacy" that wends its way through the previous passage, reaching back to Hearne's ethical foreclosure via the discourse of private property). And hence it belies Cavell's opening of the human to the animal other by rewriting the differences in *degree* in "patterns we share with other life forms" (*This New*, 48) as differences in *kind*—a maneuver made possible by grounding those otherwise conventional differences in their proper "biological" "sources." In Cavell, in other words, the opening of the human to the shared world of the animal other via the "sketchiness" of our own form of life—a sketchiness revealed in the encounter with philosophical skepticism—is in the end foreclosed by the fact that the animal other matters only insofar as it mirrors, in a diminished way, the *human* form that is the "source" of recognizing animals as bodies that have sensations, feel pain, and so on. And here Cavell's liberal humanism links him rather unexpectedly, I think, with the animal rights philosophy of Peter Singer and Tom Regan that I have already discussed, for whom our responsibility to the animal other is grounded, as I have argued, in the fact that it exhibits in diminished form qualities, potentials, or abilities that are realized to their fullest in human beings.

To put it in more strictly philosophical terms, there is a way—as Richard Rorty would no doubt be the first to argue—that all of this is already hardwired into Cavell's primary philosophical commitment to the importance of the problem of skepticism. Skepticism takes seriously, if you will, the loss of the world, its exile, as the price paid for knowledge after Kant. As Cavell writes of the Kantian "settlement" with skepticism in *In Quest of the Ordinary*, "To settle with skepticism . . . to assure us that we do know the existence of the world, or rather, that what we understand as knowledge is *of* the world, the price Kant asks us to pay is to cede any claim to know the thing in itself, to grant that human knowledge is not of things as they are in themselves. You don't—do you?—have to be a romantic to feel sometimes about that settlement: Thanks for nothing."[11] It is a "romantic" bridling against this Kantian settlement that, for Cavell, links Wittgenstein to Heidegger—*and*, as I will suggest later, opens Cavell to Derrida's critique of Heideggerian humanism. For Cavell, Wittgenstein's notion of criterion "is as if a pivot between the necessity of the relation among human beings Wittgenstein calls 'agreement in form of life' and the necessity in the relation between grammar and world," and it is this "recuperation or recoupment or

redemption of the thing (in itself)," exiled as the *Ding an sich* by Kant's
"settlement," that links Heidegger's late philosophy with Wittgenstein as "a
function of their moving in structurally similar recoils away from Kant's
settlement with the thing in itself, a recoil toward linking two 'directions' of
language—that outward, toward objects, and that inward, toward culture
and the individual" (*This New*, 49–51). For Cavell, in other words, both
Wittgenstein and Heidegger remain committed, though granted in a very
complicated way, to a fundamental alignment between the grammar of objects, of things in the world, and the grammar of language games and the
forms of life they generate; more than that, it is the biological or vertical
"form of life" of the human that is both the "source" of our attributions to
the world and the "background"—the back*ground*, to put a finer point on
it—against which they must be judged.

WHAT THE VICTIM CAN (NOT) SAY: LYOTARD (AND LÉVINAS)

However supple and nuanced the meditations on language, phenomenology, and species difference in the Wittgenstein-Cavell-Hearne line—and I
have tried to show that they are nuanced indeed—the countervaling force
of a deeply ingrained humanism in their work should impel us, I think, to
contrast their views with those of poststructuralist philosophy, because the
latter is widely held to be nothing if not posthumanist or at least antihumanist. I have in mind specifically the work of Gilles Deleuze and Félix
Guattari (to which I will turn later in this book, chiefly as background to
Michael Crichton's curious novel *Congo*), Jean-François Lyotard, and
Jacques Derrida: Lyotard, because of the tight coupling in his work of the
formal analysis of language games with questions of law and ethics and the
philosophical imperative of what he calls "the inhuman"; and Derrida, because no contemporary theorist has carried out a more searching, if epi-
sodic, investigation of the question of the animal—an investigation that
turns in no small part on an ongoing reading of Heidegger that we will soon
want to contrast with Cavell's.

For Lyotard, the question of the animal is embedded within the larger
context of the relation between postmodernity and what he has called "the
inhuman." As is well known, in *The Postmodern Condition* Lyotard borrows
the Wittgensteinian concept of the "language game" to theorize the social
and formal conditions of possibility for what he presents as a distinctly postmodern type of pluralism made possible by the delegitimizing of the "grand
metanarratives" of modernity.[12] For Lyotard, the effect of seizing on
Wittgenstein's invention is not just to radicalize his Kantian insistence on

the differences between different discourses (the descriptive and the pre-scriptive, for example) and not just to thereby "attack the legitimacy of the discourse of science" (since on this view science now "has no special calling to supervise the game of praxis"). It is also to reveal "an important current of postmodernity"—indeed, from a Lyotardian vantage, perhaps *the* most important current: that "the social subject itself seems to dissolve in this dis-semination of language games. The social bond is linguistic, but is not wo-ven with a single thread" (*Postmodern Condition*, 40). If, on this view, moder-nity consists of "a shattering of belief" and a "discovery of the 'lack of reality' of reality" (*Postmodern Condition*, 77), then what matters now is the posture one adopts toward this discovery of the postmodern at the heart of the modern:

> If it is true that modernity takes place in the withdrawal of the real . . .
> it is possible, within this relation, to distinguish two modes. . . . The
> emphasis can be placed on the powerlessness of the faculty of pre-
> sentation, on the nostalgia for presence felt by the human subject, on
> the obscure and futile will which inhabits him in spite of everything.
> The emphasis can be placed, rather, on the power of the faculty to
> conceive, on its "inhumanity" so to speak . . . on the increase of being
> and the jubilation which result from the invention of new rules of
> the game, be it pictorial, artistic, or any other. (*Postmodern Condition*,
> 79–80)

What the breakdown of the metanarratives of modernity properly calls for, then, is an opening of all language games to constant "invention" and "dis-sensus" rather than a Habermasian *con*sensus that "does violence to the het-erogeneity of language games" (*Postmodern Condition*, xxv, 65–66, 72–73); an opening to "new presentations" in the arts and literature and, in the sciences, what Lyotard calls "paralogy." This mode of scientific questioning is not reducible to the "performativity principle" of technoscience under capital but rather takes seriously such phenomena as chaos, paradox, and the like, and in so doing spurs itself toward the invention of new rules, "pro-ducing not the known but the unknown" (*Postmodern Condition*, 61).

It is against the performativity model of knowledge and legitimation and its expression in the "inhuman" juggernaut of technoscience wedded to capital (in which, as Lyotard only half jokes, "whoever is the wealthiest has the best chance of being right") that Lyotard imagines a *second* sort of "inhuman" as its antagonist. "What if human beings, in humanism's sense," he writes, "were in the process of, constrained into, becoming inhu-man. . . . [W]hat if what is 'proper' to humankind were to be inhabited by

the inhuman," a "familiar and unknown guest which is agitating it, sending
it delirious but also making it think."[13] There are, in fact, two *positive* senses
of the inhuman at work here. The first hinges on Lyotard's retheorization of
the subject as the "subject of phrases," "dispersed in clouds of narrative language elements" and components of language games, each with "pragmatic
valences specific to its kind," each giving "rise to institutions in patches—
local determinism" (*Postmodern Condition*, xxiv). This radically antianthro-
pocentric concept of the subject reaches its apotheosis in *The Differend*,
where Lyotard argues that "phrase regimes coincide neither with 'faculties
of the soul' nor with 'cognitive faculties.' . . . You don't play around with language. And in this sense, there are no language games. There are stakes tied
to genres of discourse." It is this discursive model of the subject that Lyotard
sets squarely against the "anthropocentrism" that "in general presupposes *a*
language, a language naturally at peace with itself, 'communicational' [in a
Habermasian sense], and perturbed for instance only by the wills, passions,
and intentions of humans" (137–38).[14]

Now the question squarely before us, of course, is whether this recon-
ceptualization of the subject enables us to fundamentally rethink the relations of language, ethics, and the question of the animal. In fact, Lyotard
raises this question, if only in passing, in *The Differend*—a text that seems
especially promising in this connection in its resolute antianthropocen-
trism:

> French *Aie*, Italian *Eh*, American *Whoops* are phrases. A wink, a shrug-
> ging of the shoulder, a taping [*sic*] of the foot, a fleeting blush, or an at-
> tack of tachycardia can be phrases.—And the wagging of a dog's tail,
> the perked ears of a cat?—And a tiny speck to the West rising upon the
> horizon of the sea?—A silence? . . . —Silence as a phrase. The expec-
> tant wait of the *Is it happening?* as silence. Feelings as a phrase for what
> cannot now be phrased. (70)

Here Lyotard seems to extend the sense of "language games" in his earlier
work, via a rather capacious concept of the "phrase," in directions not unlike
those Hearne develops in her work on transspecies communication.

And this possibility is only further strengthened by the introduction to
the essays collected in *The Inhuman*, where he offers a gloss on the inhuman
that is worth quoting at length:

> What shall we call human in humans, the initial misery of their child-
> hood, or their capacity to acquire a "second" nature which, thanks to
> language, makes them fit to share in communal life, adult conscious-

ness and reason? That the second depends on and presupposes the first is agreed by everyone. The question is only that of knowing whether this dialectic, whatever name we grace it with, leaves no remainder.

If this were the case, it would be inexplicable for the adult himself or herself not only that s/he has to struggle constantly to assure his or her conformity to institutions . . . but that the power of criticizing them, the pain of supporting them and the temptation to escape them persist in some of his or her activities. . . . There too, it is a matter of traces of indetermination, a childhood, persisting up to the age of adulthood.

It is a consequence of these banal observations that one can take pride in the title of humanity, for exactly opposite reasons. Shorn of speech, incapable of standing upright, hesitating over the objects of interest, not able to calculate its advantages, not sensitive to common reason, the child is eminently human because its distress heralds and promises things possible. Its initial delay in humanity, which makes it the hostage to the adult community, is also what manifests to this community the lack of humanity it is suffering from, and which calls on it to become more human. (3–4)

It is not enough—and here we can cast a glance backward at Luc Ferry's formulation of humanism in chapter 1—that "our contemporaries find it adequate to remind us that what is proper to humankind is its absence of defining property, its nothingness, or its transcendence, to display the sign 'no vacancy,'" for what such a posture "hurries, and crushes, is what after the fact I find I have always tried, under diverse headings—work, figural, heterogeneity, dissensus, event, thing—to reserve: the unharmonizable" (4). The child, then, inhabits the inhuman in the same way that the postmodern inhabits the modern, and what makes this analogy initially seem so useful for theorizing the animal other is that it posits a permanently incipient multiplicity and self-difference at the very core of subjectivity as such, and in doing so promises to help us extend contemporary transvaluations of the structural homology between child and animal available to us at least since Freud. I will discuss these homologies later at several junctures, including the psychoanalytic reworking of Lacan, Hegel, and Kant that we find in Slavoj Žižek's work on "the Thing" and monstrosity (key to our understanding of the film *The Silence of the Lambs*), and the *anti*psychoanalytic work of Deleuze and Guattari on "becoming animal" (which will be crucial to the analysis of the novel *Congo*).[15]

Lyotard's work thus seems at first to mark an advance beyond Cavell's on the question of the animal. For both—and for both within a Kantian frame of sorts—the animal marks an outside or limit that is of a piece with the Kantian Thing, in the face of which knowledge comes to an end. And in and by that end, the ends of the humanist model of subjectivity are interrogated. Unlike Cavell's skepticism, however, Lyotard does not regard this "withdrawal of reality" nostalgically, as a "loss" of reality, but rather finds in it a generative possibility for pluralism. More pointedly, and in more strictly philosophical terms, Lyotard does not retain nostalgia, as Cavell's skeptical frame does, for some representational alignment, however sophisticated, between the grammar of language games and the grammar of the world of objects—a nostalgia that becomes problematic, as we have seen, in Cavell's reading of the "biological" sense of Wittgenstein's "form of life."

In Lyotard, however, this potential opening for theorizing the standing of the animal other is foreclosed, in the end, by the very Kantianism he shares with Cavell. As he explains early in *The Differend*—in a passage we should hear in concert with the earlier quotation on the dog's tail, the cat's perked ears, and "silence as a phrase,"

> The differend is the unstable state and instant of language wherein something which must be able to be put into phrases cannot yet be. This state includes silence, which is a negative phrase, but it also calls upon phrases which are in principle possible. This state is signaled by what one ordinarily calls a feeling. . . . In the differend, something "asks" to be put into phrases, and suffers from the wrong of not being able to be put into phrases right away. This is when the human beings who thought they could use language as an instrument of communication learn through the feeling of pain which accompanies silence (and of pleasure which accompanies the invention of a new idiom), that they are summoned by language, . . . that what remains to be phrased exceeds what they can presently phrase, and that they must be allowed to institute idioms which do not yet exist. (13)

What bars the animal from this otherwise potentially welcoming theorization is the direct linkage in Lyotard between the "feeling" of something that "asks" to be phrased and the Kantian notions of the presentable and the sublime that Lyotard develops in a number of texts. As he had already explained in *The Postmodern Condition*, the "strong and equivocal emotion" of the sublime sentiment is indicative of the "conflict between the faculties of a subject, the faculty to conceive something and the faculty to 'present' something" (77). And it takes place "when the imagination fails to present an object which might, if only in principle, come to match a concept. We have the Idea of the world (the totality, of what is) but we do not have the capacity to show an example of it"—such Ideas are "unpresentable" (78). It is the sublime sentiment, born of this conflict, that creates differends and is the spur for new phrases, new discursive rules, and inventions.

That the Kantian problematic of the sublime provides the overarching context for the earlier passage I quoted on "feeling," "silence," and animal kinesics in relation to phrases is even clearer in *The Differend*. And the problem is that once these "silences" and "emotions" are framed in Kantian terms, a certain order of subject is presupposed that automatically prevents the animal from occupying any of the discursive positions necessary for the ethical force of the differend to apply. The "silence" and "feeling" of the mute or unspoken are not available to the animal, because animals do not possess the capacity to phrase; thus their silence and feeling, even if they can be said to exist, cannot express a differend; it is not a withholding, and thus it does not express the ethical imperative of dissensus and the differend. As Lyotard writes in *Just Gaming* of the ethical call, the position of the addressee is privileged: "First, one acts from the obligation that comes from the simple fact that I am being spoken to, that you are speaking to me, and then, and only then, can one try to understand what has been received. In other words, the obligation operator comes first and then one sees what one is obligated to."[16] In this sense, as he explains, ethics has no positive content. "There is no content to the law," Lyotard writes. "And if there is no content, it is precisely because freedom is not determinant. Freedom is regulatory; it appears in the statement of the law only as that which must be respected; but one must always reflect in order to know if in repaying a loan or in refusing to give away a friend, etc., one is actually acting, *in every single instance*, in such a way as to maintain the Idea of a society of free beings" (*Just Gaming*, 85). The famous "so that" (*so dass*) of Kant's categorical imperative "does not say: 'If you want to be this, then do that'" but rather "marks the properly reflective use of judgment. It says: Do whatever, not on condition that, but *in such a way as* that which you do, the maxim of what you do, can always be valid as, etc. We are not dealing here with a determinant synthesis but with an Idea of human society" (*Just Gaming*, 85).

Here the linkage between a particular notion of the subject and a specific sense of ethics is very close to what we find in the work of Emmanuel Lévinas—a connection that seems to have reached its high-water mark in Lyotard's work during the period of the conversations with Jean-Loup Thébaud collected under the title *Just Gaming*.[17] There Lyotard explains that it is "the absolute privileging of the pole of the addressee" in Lévinas that "marks the place where something is prescribed to me, that is, where I am obligated before any freedom" (37). This means that the ethical "you must," the obligation attendant on the addressee, the prescriptive as such, cannot be "derived" from reason (or in Kantian terms, from the descriptive). And so it is folly—and in Lyotard's terms, in fact, a form of terrorism—to try to offer reasons for the origin or content of ethical obligation. "The 'you must,'" Lyotard writes, "is an obligation that ultimately is not even directly experienced"; it "is something that exceeds all experience" (45–46).[18]

The question, then, is whether this Lévinasian sense of the ethical makes it possible to rethink the question of the nonhuman animal. John Llewelyn, in a concise and exacting essay titled "Am I Obsessed by Bobby? (Humanism of the Other Animal)," has tackled this question head-on. Bobby (as the more dedicated readers of Lévinas will know) is the name of a dog Lévinas writes about in an essay from 1975 in which, as Llewelyn puts it, he "all but proposes an analogy between the unspeakable human Holocaust and the unspoken animal one."[19] Bobby strayed into the prison camp where Lévinas and his fellow Jewish prisoners had themselves "become accustomed to being treated as less than human" (235) and offered, as dogs will do, friendship and loyalty to the prisoners, greeting them at the end of each day with bright eyes and wagging tail without regard for their "inhuman" condition. But the problem for Lévinas, according the Llewelyn, is that "Bobby lacks the brains to universalize his maxim. He is too stupid, *trop bête*. Bobby is without *logos* and that is why he is without ethics . . . since the ethics of Emmanuel Lévinas is analogous to the ethics of Immanuel Kant." As Kant writes, "Since in all our experience we are acquainted with no being which might be capable of obligation (active or passive) except man, man therefore can have no duty to any being other than man" (qtd. in Llewelyn, 236). As Llewelyn takes pains to point out, it is not that the question famously raised by Jeremy Bentham—Can animals suffer?—is irrelevant for Kant.[20] If in Kant's view we seek our own happiness as a "natural end," and "since that natural end includes man's well-being as an animal, the maxim 'Treat nonhuman animals as if they have no capacity for suffering' is not one that can be consistently conceived as a law of nature," because "such a conception is inconsistent with what one knows about animals from one's own experience of being one" (241).

At the same time, however, Kant "remains adamant that we can have direct duties only to beings that have *Wille* understood as pure practical reason" (241). And for Lévinas, according to Llewelyn, things are even more stringent than for Kant. First, it is crucial to Lévinas "whether in the eyes of the animal we can discern a recognition, however obscure, of his own mortality . . . whether, in Lévinas' sense of the word, the animal has a face" (240). For only if it does can the ethical call of "the first word addressed to me by the Other"—"Thou shalt not murder/kill"—apply to my relation with a nonhuman other. And here, for Lévinas, the answer is quite unambiguously no (243). Second, for Lévinas "I can have direct responsibilities only toward beings that can speak"; both Lévinas and Kant (like Hearne) "require an obligating being to be able to make a claim in so many words. No claim goes without saying, even if the saying is the silent saying of the discourse of the face"—a formulation that ratifies Lyotard's Kantian reading of "feelings," "silence," and the "withholding" of the phrase that in the end excludes the animal in *The Differend*. In an echo of Cavell's meditation on "the romance of the hand and its apposable thumb," "the upright posture" and "the eyes set for heaven," we find in Lévinas that "the Other has only to look at me. Indeed, what is expressed in his face may be expressed by his hand or the nape of his neck" (241)—the full resonance of which I will explore in a moment in Derrida's reading of "Heidegger's Hand." And though for Lévinas this "very *droiture* of the face-to-face, its uprightness or rectitude, is the expression of the other's *droit* over me," that relationship can never include Bobby or any animal who, deprived of *Wille*, reason, and language, remains, for all ethical purposes, faceless (Llewelyn, 242).

Similarly, in Lyotard one does not know what the ethical call calls for, but one certainly knows *whom* it calls for:

> There is a willing. What this will wants, we do not know. We feel it in the form of an obligation, but this obligation is empty, in a way. So if it can be given a content in the specific occasion, this content can only be circumscribed by an Idea. The Idea is . . . "the whole of reasonable beings" or the preservation of the possibility of the prescriptive game. But this whole of reasonable beings, I do not know if the will wants it or what it wants with it. I will never know it. (*Just Gaming*, 70)[21]

Lyotard's answer to the question he poses in *The Differend*—"The wagging of a dog's tail, the perked ears of a cat?"—will come as no surprise, then, when he writes earlier in the book that the animal, because it does not have the means to bear witness, is "a paradigm of the victim" (28) who suffers wrongs but cannot claim damages:

> Some feel more grief over damages inflicted upon an animal than over those inflicted upon a human. This is because the animal is deprived of the possibility of bearing witness according to the human rules for

establishing damages, and as a consequence, every damage is like a wrong and turns it into a victim *ipso facto.*—But, if it does not at all have the means to bear witness, then there are not even damages, or at least you cannot establish them. . . . That is why the animal is a paradigm of the victim. (28)

Thus we are returned in Lyotard's work, via Kant, to an essential (if extremely sophisticated) humanism regarding the ethical and the animal: first, in the taken-for-granted muteness of the animal, which, crucially, can never be a "withholding" that, via the "feelings" that generate differends, is ethically productive of or included in the postmodern pluralism that Lyotard wants to promote; and second, in the theorizing of the ethical community of "reasonable beings" whose standing is grounded in the capacity for language, whether formalized subsequently by the social contract to which only humans are party or by the reinstatement of the Kantian divide between direct duty to humans and indirect duty to animals. For Lyotard as for Cavell, it is on the specific site of the ethical standing of the animal other that we get the clearest picture of a humanism that is otherwise sometimes hard to see. For both, the animal is that Kantian outside that reveals our traditional pictures of the ontological fullness of the human to have been fantasies all along, built on the sands of disavowal of our own contingency, our own materiality, our own "spokenness." But once that work is done, the animal is returned to its exile, its facelessness, as the human now retains a privileged relation—indeed a constitutive one—not to its own success but to its hard-won failure, from which the animal remains excluded. In the end, for Lyotard, we may not be us, but at least we retain the certainty that the animal remains the animal.

"THE ANIMAL, WHAT A WORD!" DERRIDA (AND LÉVINAS)

Given the shortcomings of the Lyotardian frame, I turn now to the work of Jacques Derrida, who writes in *Of Spirit: Heidegger and the Question* that the "discourse of animality remains for me a very old anxiety, a still lively suspicion," one to which he has "made numerous references" over "a very long period."[22] This is certainly true, but this attention to the question of the animal seems to have reached a new pitch of intensity and, one is tempted to say, passion or *com*passion in Derrida's recent work, most notably the essay "The Animal That Therefore I Am (More to Follow)," delivered as the opening part of a ten-hour lecture at Cerisy-la-Salle in 1997 at a conference devoted to Derrida's work, titled "L'animal autobiographique." There he lists upward of twenty texts in which the question of the animal has arisen throughout his career—and nowhere more densely, perhaps, than in his reading of the figure of "the hand" in relation to the human/animal duality in his work on Heidegger.

In "*Geschlecht* II: Heidegger's Hand," Derrida makes a statement that must seem to any reader—perhaps *especially* to those who think of themselves as Derrideans—a sweeping one indeed, when he writes of Heidegger's work on the hand that "here in effect occurs a sentence that at bottom seems to me Heidegger's most significant, symptomatic, and seriously dogmatic," one that risks "compromising the whole force and necessity of the discourse." The sentence he has in mind from Heidegger is this: "Apes, *for example* [my emphasis, J. D.], have organs that can grasp, but they have no hand."[23] What can Heidegger mean here, particularly since such a statement remains, as Derrida notes, willfully ignorant of the whole body of "zoological knowledge" to the contrary (173)? First, we should remember that Heidegger's larger political interest in thinking of the meaning of *Geschlecht* (the *genre humaine* or species being, "the humanity of man" [163])—an altogether understandable one, as Derrida notes—is to "distinguish between the national and nationalism, that is, between the national and a biologicist and racist ideology" (165).[24]

What Heidegger has in mind, then, is a figure of the hand whose being is determined not by biological or utilitarian function—that "does not let itself be determined as a bodily organ of gripping" (172)—but rather can serve as a figure for thought, and a particular mode of thought at that, that distinguishes the *Geschlecht* of humanity from the rest of creation. "If there is a thought of the hand or a hand of thought, as Heidegger gives us to think," Derrida writes, "it is not of the order of conceptual grasping. Rather this thought of the hand belongs to the essence of the *gift*, of a giving that would give, if this is possible, without taking hold of anything" (173). We find here a contrast—an "abyss" in fact, as Derrida will argue—between the grasping or "prehension" associated with the "prehensile" organs of the ape (*Of Spirit*, 11) and the hand of man, which "is far from these in an infinite way (*unendlich*) through the abyss of its being. . . . This abyss is speech and thought. 'Only a being who can speak, that is, think,'" Heidegger writes, "'can have the hand and be handy (*in der Handhebung*) in achieving works of handicraft.'" "The hand," Heidegger writes, "does not only grasp and catch (*greift und fangt nicht nur*). The hand reaches and extends, receives and welcomes . . . extends itself, and receives its own welcome in the hand of the other" (qtd. in "*Geschlecht* II," 174). Add to this Heidegger's contention a page later: "Only when man speaks does he think—not the other way around, as metaphysics still believes. Every motion of the hand in every one of its works carries itself (*sich tragt*) through the element of thinking, every bearing of the hand bears itself (*gebardet sich*) in that element. All the work of the hand is rooted in thinking. Therefore, thinking (*das Denken*) itself is man's simplest, and for that reason hardest, *Hand-Werk*" (qtd. in "*Geschlecht* II," 175).

We should be reminded here, I think, of a similar moment in Cavell's reading of Heidegger that takes the statement "Thinking is a handicraft" to mean not just that the hand and the "fantasy of the apposable thumb" figure thought as a distinctly human relation to the world, but also that they figure Heidegger's "interpretation of Western conceptualizing as a kind of sublimized violence," a sort of "clutching" or "grasping" through what we might call "prehensile" conceptualization—a mode of violence famously thematized in Heidegger as the violence "expressed in the world dominion of technology" (*Conditions Handsome*, 38, 41).[25] In opposition to all of which Cavell finds Heidegger's emphasis on thought as "reception," as a kind of welcoming, elaborated by Heidegger in passages that insist on "the derivation of the word thinking from a root for thanking and interprets this particularly as giving thanks for the gift of thinking" (38–39).

It should not surprise us at this juncture that Derrida's critique of this cluster of figures in Heidegger is surely more pointed than Cavell's, since Cavell, as we have seen, remains in some important sense a part of that humanist tradition to which Heidegger belongs. Or to put it another way, Cavell's taking seriously the problem of skepticism is simultaneously taking seriously the nondeconstructability of the opposition between giving and taking. But "the nerve of the argument," Derrida writes, "seems to me reducible to the assured opposition of *giving* and *taking*: man's hand *gives and gives itself, gives* and *is given*, like thought . . . whereas the organ of the ape or of man as a simple animal, indeed as *animal rationale*, can only *take hold of, grasp, lay hands on the thing*. The organ can *only* take hold of and manipulate the thing insofar as, in any case, it does not have to deal with the thing *as such*, does not let the thing be what it is in its essence" ("*Geschlecht* II," 175). But of course—and here is the difference with Cavell—"nothing is less assured," as Derrida has argued in any number of texts, "than the distinction between *giving* and *taking*, at once in the Indo-European languages we speak . . . and in the experience of an economy" (176).

Heidegger's hand is only an especially dense and charged figure for what Derrida in *Of Spirit* will critique in Heidegger as "the profoundest metaphysical humanism," subjecting to rigorous deconstruction Heidegger's tortured theses in *Fundamental Concepts of Metaphysics* that (1) "the stone is without world," but (2) "the animal is poor in world," unlike (3) man, who is "world-forming" or world building (48). As Derrida remarks, what at first looks like a difference only in *degree* between the "poverty" of the animal and the plenitude of the human in relation to having a world is paradoxically maintained by Heidegger as a difference in *kind*, a "difference in essence" (48–49). The central problem here—it is one entirely symptomatic of Heidegger's humanist project—is one of "two values incompatible in their 'logic': that of lack and that of alterity" (49); in the interests of determining the "we" of *Dasein*, of Being, "The lack of world for the animal is not a pure nothingness"—as it would be for the stone—"but it must not be referred, on a scale of homogeneous degrees, to a plenitude, or to a non-lack in a heterogeneous order, for example that of man" (49). The animal for Heidegger, as Derrida characterizes it, therefore paradoxically "has a world in the mode of not-having" (50); it "can have a world because it has access to entities, but it is deprived of a world because it does not have access to entities *as such* and in their Being" (51). The lizard stretched on the rock in the sun, as Heidegger puts it in a famous example, cannot relate to the rock and sun *as such*, "as that with regard to which, precisely, one can put questions and give replies" (52), and this is so because the lizard does not have language. As Derrida emphasizes, "This inability to name is not primarily or simply linguistic; it derives from the properly *phenomenological* impossibility of speaking the phenomenon whose phenomenality as such, or whose very *as such*, does not appear to the animal and does not unveil the Being of the entity" (53). For Heidegger, then, "There is no animal *Dasein*, since *Dasein* is characterized by access to the 'as such' of the entity and to the correlative possibility of questioning." The animal has no hand, or to put it in the Lévinasian terms that I will contrast with Derrida shortly, the animal has no face; it cannot be an other.

A fundamental symptom (or depending on how one reads Derrida, I suppose, the fundamental cause) of this rhetoric of the animal in Heidegger that brings "the consequences of a serious mortgaging to weigh upon the whole of his thought" (*Of Spirit*, 57) is that it is presented in the dogmatic form of a *thesis*—a reductive genre that Derrida clearly bridles against in principle. The form of thesis presupposes "that there is one thing, one domain, one homogeneous type of entity, which is called animality *in general*, for which any example would do the job" (57). The monstrosity of the thesis is, in a word, its dogmatism, and it partakes of the same logic that drives the "monstrosity" of Heidegger's hand. This monstrosity becomes for Derrida a figure for Heidegger's flight from *différance* generally, but specifically as it is disseminated through the sites of species difference and sexual difference—a double point that will help make clear Derrida's differences with Lévinas as well. As Derrida characterizes this "monstrosity" called man, "*The* hand of *the* man, of man *as such;* . . . Heidegger does not only think of the hand as a very singular thing that would rightfully belong only to man, he always thinks the hand *in the singular,* as if man did not have two hands but, this monster, one single hand" ("*Geschlecht* II," 182).

It is the rejection of "animality in general," and of singularity and identity *in general,* that is amplified considerably in Derrida's recent lecture "The Animal That Therefore I Am (More to Follow)." The "animal, what a word!" Derrida writes (48):

> Confined within this catch-all concept, within this vast encampment of the animal, in this general singular, within the strict enclosure of this definite article ("the Animal" and not "animals") . . . are *all the living things* that man does not recognize as his fellows, his neighbors or his brothers. And that is so in spite of the infinite space that separates the lizard from the dog, the protozoon from the dolphin, the shark from the lamb, the parrot from the chimpanzee. (51)

For Derrida, this "immense multiplicity of other living things . . . cannot in any way be homogenized, except by means of violence and willful ignorance" (72). "The confusion of all non-human living creatures within the general and common category of the animal is not simply a sin against rigorous thinking, vigilance, lucidity or empirical authority," he continues, "it is also a crime. Not a crime against animality precisely, but a crime of the first order against the animals, against animals. Do we agree to presume that every murder, every transgression of the commandment 'Thou shalt not kill' concerns only man?" (73). Here, of course, Derrida offers a reprise of the diagnosis of the "carnophallogocentrism" of the Western philosophical tradition that he discusses at some length in "'Eating Well.'" In both texts the Word, *logos,* does violence to the heterogeneous multiplicity of the living world by reconstituting it under the sign of identity, the *as such* and *in general*—not "animals" but "*the* animal." And as such, it enacts what Derrida calls the "sacrificial structure" that opens a space for the "noncriminal putting to death" of the animal—a sacrifice that (so the story of Western philosophy goes) allows the transcendence of the human, of what Heidegger calls "spirit," by the killing off and disavowal of the animal, the bodily, the materially heterogeneous, the contingent—in short, of *différance* ("'Eating Well,'" 113).

And yet Derrida's recent work moves beyond that analysis, or perhaps fleshes out its full implications (if you'll allow the expression), in a couple of very important ways—ways that will, moreover, sharpen our sense of Derrida's important relationship with Lévinas on the question of ethics. For in the Cerisy lecture Derrida is struggling to say, I believe, that the animal difference is, *at this very moment*, not just any difference among others; it is, we might say, the most different difference, and therefore the most instructive—*particularly* if we pay attention, as he does here, to how it has been consistently repressed even by contemporary thinkers as otherwise profound as Lévinas and Lacan. To pay proper attention to these questions, "it would not be a matter of 'giving speech back' to animals," Derrida writes, "but perhaps of acceding to a thinking, however fabulous and chimerical it might be, that thinks the absence of the name and of the word otherwise, as something other than a privation" ("The Animal That Therefore I Am," 73). It would be to enact, as it were, a radical transvaluation of the "reticence" of Wittgenstein's lion. But how to do this?

In a move that is bound to be surprising, I think, Derrida returns to the central question famously raised by Jeremy Bentham in response to Descartes: The question with animals is not can they talk, or can they reason, but can they *suffer*. "Once its protocol is established," Derrida writes, "the form of this question changes everything," because "from Aristotle to Descartes, from Descartes, especially, to Heidegger, Lévinas and Lacan," posing the question of the animal other in terms of *logos*, of either thought or language, "determines so many others concerning *power* or *capability* [*pouvoirs*], and *attributes* [*avoirs*]: being able, having the power to give, to die, to bury one's head, to dress, to work, to invent a technique" (41). What makes Bentham's reframing of the question so powerful is that now "the question is disturbed by a certain *passivity*. It bears witness, manifesting already, as question, the response that testifies to sufferance, a passion, a not-being-able." "What of the vulnerability felt on the basis of this inability?" he continues. "What is this non-power at the heart of power? . . . What right should be accorded it? To what extent does it concern us?" (42). It concerns us very directly, in fact—as we know from both Heidegger and Lévinas—for "mortality resides there, as the most radical means of thinking the finitude that we share with animals, the mortality that belongs to the very finitude of life, to the experience of compassion, to the possibility of sharing the possibility of this non-power, the possibility of this impossibility, the anguish of this vulnerability and the vulnerability of this anguish" (42).[26]

It is here, at this precise juncture, that Derrida's links and differences with Lévinas—and for that matter with Lyotard—become most pronounced and most pointed. On the one hand, they share a certain sense of ethics. As Richard Beardsworth explains in *Derrida and the Political*, the relation between ethics, the other, and time is central to the critique of Heidegger in both Derrida and Lévinas. For both, "Time is not only irrecoverable; being irrecoverable, time is ethics."[27] Even more to the point for the "passivity" and "vulnerability" of the animal other invoked by Derrida is that Heidegger *appropriates* the limit of death "rather than returning it to *the other* of time. The existential of 'being-towards-death' is consequently a 'being-able' (*pouvoir-être*), not the impossibility of all power." For Lévinas and Derrida, on the other hand,

> the "impossibility" of death for the ego confirms that the experience of finitude is one of radical passivity. That the "I" cannot experience its "own" death means, firstly, that death is an immanence *without* horizon, and secondly, that time is that which exceeds my death, that time is the generation which precedes and follows me. . . . Death is not a limit or horizon which, re-cognized, allows the ego to assume the "there" [as in Heidegger's "being-toward-death"]; it is something that never arrives in the ego's time, a "not-yet" which confirms the priority of time over the ego, marking, accordingly, the precedence of the other over the ego. (Beardsworth, 130–31)

What this means is that "death *im*possibilizes existence," and does so both for me *and* for the other—since death is no more "for" the other than it is for me—so that "the alterity of death rather than signalling the other signals the *alterity* of the other, the other, if one wishes, as the recurrence of time" (132).

For Lévinas and for Derrida, this has crucial implications for their view of ethics, because it suggests that the subject is always "too late" in relation to the other qua the absolute past, even as it is in that relation that the ethical fundamentally resides. At the root of ethical responsibility, then, is paradoxically its impossibility. But it is in this impossibility that the possibility of justice resides—a justice not reducible to the immanence of any particular socially or historically inscribed doctrine of law. As Derrida explains in "Force of Law: The 'Mystical Foundation of Authority,'"

> A decision that did not go through the ordeal of the undecidable would not be a free decision, it would only be the programmable application of unfolding of a calculable process. It might be legal; it would not be just. . . . Here we "touch" without touching this extraordinary paradox: the inaccessible transcendence of the law before which and prior to which "man" stands fast only appears infinitely transcendent and thus theological to the extent that, so near him, it

depends only on him, on the performative act by which he institutes it." (Qtd. in Beardsworth, 44–45)

And it is here, of course, that the sense of ethics in Lévinas and Derrida is diametrically opposed to what we find in a utilitarian like Peter Singer, the leading figure in animal rights philosophy. For in Singer, as we have already seen, ethics means precisely the application of a "calculable process," namely, the utilitarian calculus that would tally up the "interests" of the particular beings in question in a given situation, regardless of their species, and would determine what counts as a just act according to which action maximizes the greatest good for the greatest number. In doing so, however, Singer's utilitarian ethics would violate everything that the possibility of justice depends on in Derrida. First, it would run aground on Kant's separation of prescriptive and descriptive discourses, because "if one knew how to be moral, if one knew how to be free, then morality and freedom would be objects of science" (Beardsworth, 52)—and we all know that there is no science of ethics. Second and more seriously—and Derrida is quite forceful on this point—it reduces ethics to the very antithesis of ethics by reducing the aporia of judgment in which the possibility of justice resides to the mechanical unfolding of a positivist calculation. This is what Derrida has in mind, I think, when he writes,

> I have thus never believed in some homogeneous continuity between what calls *itself* man and what *he* calls the animal. I am not about to begin to do so now. That would be worse than sleepwalking, it would simply be too asinine [*bête*]. To suppose such a stupid memory lapse or to take to task such a naïve misapprehension of this abyssal rupture would mean, more seriously still, venturing to say almost anything at all for the cause. . . . When that cause or interest begins to profit from what it simplistically suspects to be a biological continuism, whose sinister connotations we are well aware of, or more generally to profit from what is suspected as a geneticism that one might wish to associate with this scatterbrained accusation of continuism, the undertaking becomes aberrant. ("The Animal That Therefore I Am," 45–46)

From Derrida's point of view, then, the irony of Singer's utilitarian calculus, even if it is in the service of "the cause" of the animal, is that it would be "asinine"—bestial in fact—not only because of its "geneticism" and "continuism" (most clearly in its concept of "interests") but also because it would be, ironically enough, the sort of application of a mechanical program of behavior by an automaton that Descartes associated with the animal and the "bestial."[28]

This does not mean, of course, that Derrida does not take seriously the ethical question of nonhuman animals or, for that matter, all the issues associated with the term "animal rights." Indeed, it is this as much as anything that separates him from Lévinas. I have already touched on this, but here I could do no better than to recall Derrida's own discussion of Lévinas's attractions and limits in "'Eating Well.'" For Lévinas, subjectivity "is constituted first of all as the subjectivity of the *hostage*"; the subject is held hostage by the other, in responsibility to the other, in the imperative "Thou shalt not kill." But in Lévinas, as in the Judeo-Christian tradition generally, this is not understood as "Thou shalt not put to death the living in general" (112–13). But why not? Because, as Derrida shows, "Lévinas's thematization of the other 'as' other presupposes the 'as'-structure of Heideggerian ontology" (Beardsworth, 134). It holds, that is, that the other can appear *as such*—not as an ontological positivity, as in Heidegger, but rather as a form of *privileged negativity* (what Lévinas often calls "passivity" or "anarchy" or "vulnerability") that is *always* the form of the ethical *as such*. For Derrida, on the other hand, one must keep the "there" of ethics, the site of the other, "as complex as possible, as a 'play' of time and law, one which refuses the exemplary localization of thought" of the sort that we find in Lévinas's contention that the "authentically human" is the "being-Jewish in every man" (Beardsworth, 124).

But for Derrida, "For the other to be other it must already be less than other," because the alterity of the other is always already caught in what Derrida in "'Eating Well'" calls the "sacrificial economy" of carnophallogocentrism; hence "one cannot 'welcome the other as other.'" In consequence of which, as Beardsworth notes, "alterity can only be the loss of the other in its self-presentation, that is, the 'trace' of the other" (134). What Lévinas surrenders, then, is "a differentiated articulation *between* the other and the same," the effect of which "is the loss in turn of the *incalculable* nature of the relation between the other and its others (the community at large)" (Beardsworth, 125).

For Derrida—to return to "'Eating Well'"—the surest sign of this recontainment of the alterity of the other in Lévinas is that the ethical status of the "community at large" is purchased at the expense of the sacrifice of all forms of difference that are not human—most pointedly, of course, the animal—whereas for Derrida the animal *in the plural* is precisely what keeps open the ethical moment of the self via its passivity because the animal's *death, its mortality, is not sacrificed.* "Discourses as original as those of

Heidegger and Lévinas, disrupt, of course, a certain traditional humanism," Derrida holds in "'Eating Well.'" "In spite of the differences separating them, they nonetheless remain profound humanisms *to the extent that they do not sacrifice sacrifice*. The subject (in Lévinas's sense) and the *Dasein* are 'men' in a world where sacrifice is possible and where it is not forbidden to make an attempt on life in general, but only on the life of man" (113). For Derrida, on the other hand, the animal "has its point of view regarding me. The point of view of the absolute other, and nothing will have ever done more to make me think through this absolute alterity of the neighbor than these moments when I see myself seen naked under the gaze of a cat" ("The Animal That Therefore I Am," 16).

And when Derrida says "man" here we should, I think, hear him quite pointedly, for the problem with animal difference is strictly analogous to the recontainment of *sexual* difference in both Heidegger and Lévinas.[29] As for the latter, Derrida explains that from Lévinas's point of view it is not woman or femininity per se but rather sexual difference itself that is ethically secondary, the point being that "the possibility of ethics could be saved, if one takes ethics to mean that relationship to the other as other which accounts for no other determination or sexual characteristic in particular. What kind of an ethics would there be if belonging to one sex or another became its law or privilege?" And yet, Derrida continues, it is not clear that Lévinas is not here restoring "a classical interpretation" that "gives a masculine sexual marking to what is presented either as a neutral originariness or, at least, as prior and superior to all sexual markings . . . by placing (differentiated) sexuality beneath humanity which sustains itself at the level of Spirit."[30] And that "humanity," in turn, depends on the sacrificial structure that orders the relationship between the world "of spirit" and the animal. Hence the full force of Derrida's comment late in the Cerisy lecture that, in the philosophical tradition, he has never "noticed a protestation *of principle*, and especially a protestation of consequence against the general singular that is the *animal*. Nor against the general singular of an animal whose sexuality is as a matter of principle left undifferentiated—or neutralized not to say castrated" (61).

If Derrida's differences with Lévinas on ethics, writing, and the animal are perhaps clear by now, it is worth briefly examining his differences with Lyotard as well. All three share the sense of ethics voiced in Lyotard's *Just Gaming*: that "any attempt to state the law, for example, to place oneself in the position of enunciator of the universal prescription is obviously infatuation itself and absolute injustice, in point of fact. And so, when the question of what justice consists in is raised, the answer is: 'It remains to be seen in each case'" (99). But Derrida would draw our attention to the ethical stakes involved for "'the crossing of borders' between man and animal" ("The Animal That Therefore I Am, 4") in their respective theories of language, writing, the phrase, and so on. Here what we might call Lyotard's radical formalism appears to be problematic, for as Sam Weber notes in his afterword to *Just Gaming*, in Lyotard "the concern with 'preserving the purity' and singularity 'of each game' by reinforcing its isolation from the others gives rise to exactly what was intended to be avoided: 'the domination of one game by another, namely, 'the domination of the prescriptive,'" in the form of *Thou shalt not let one language game impinge on the singularity of another* (104). To put it another way, if in Lyotard the Kantian "outside" marked by the difference between the conceivable and the presentable is what permanently keeps open the ethical necessity of dissensus and invention, the price Lyotard pays for this way of formulating the problem is that the language games themselves become in an important sense pure and self-identical, and hence the boundaries between them become in principle absolutely uncrossable. Thus the field of "general agonistics" of which, for Lyotard, any language game partakes (*Postmodern Condition*, 10) is, as Weber rightly points out, not so agonistic, or so general, after all, since it is restricted by the countervailing force of Lyotard's concept of the language game, which can be in struggle neither internally (since it is a singularity determined by a finite set of rules) nor externally (since the incommensurability of all games is to be protected at all costs) (*Just Gaming*, 104).

For Derrida, on the other hand, the outside is always already inside; in Lyotardian terms, the verticality of the language game is always already constitutively eroded by the horizontality of the field of inscription and signification—of *différance* and the trace of writing—of which it is part. And hence the ethical subject of the Kantian "Idea" in Lyotard's scheme—the subject of the "community of reasonable beings"—is always already constitutively derailed by the *un*reason, the *a*logological force of the *écriture* on whose disavowal the Law constructs itself in a process that Derrida calls "the law of Law." For Kant, we should remember, "the moral law is transcendent because it transcends the sensible conditions of time and space." But for Derrida, the *différance* of law, the law of Law, consists in the fact that "if the law is, on the one hand, unaccountable" (and this is where Derrida's relationship with Lévinas is triangulated via different relations to Kant), "on the other hand it is *nowhere* but *in* its inscriptions in history, whilst not being reducible *to* these inscriptions either" (Beardsworth, 29).

Thus the Kantian gives way to the Nietzschean realization, as Weber puts it, that "otherness, then, is not to be sought *between* games that are supposed to be essentially self-identical, but *within* the game as such" (*Just*

Gaming, 106). Or as Geoffrey Bennington characterizes it in more strictly Derridean terms, for Derrida "language is not essentially human . . . ; the refusal to think of language as in some way a separate domain over against the world . . . implies the consequence of an essential inhumanity of language."[31] And this difference between Lyotard's sense of language and Derrida's has very direct implications for conceptualizing the problematic of the animal in relation to ethics. As Vicki Kirby points out, if one

> reads the substance of materiality, corporeality, and radical alterity together, and places them outside or beyond representation, the absolute cut of this division actually severs the possibility of an ethical relation with the Other. . . . [E]thical responsibility to the Other therefore becomes an act of conscious humility and benevolent obligation to an Other who is not me, an Other whose difference is so foreign that it cannot be known. Yet a Derridean reading would surely discover that the breach in the identity and being of the sovereign subject, and in the very notion of cognition itself, is not merely nostalgic loss nor anticipated threat or promise. It is a constitutive breaching, a recalling and differentiating within the subject, that hails it into presence. As impossible as it may seem, the ethical relation to radical alterity is to an other that is, also, me. (95)

This is precisely what Derrida has in mind, I think, when he contends in "'Eating Well'" that

> the idea according to which man is the only speaking being, in its traditional form or in its Heideggerian form, seems to me at once undisplaceable and highly problematic. Of course, if one defines language in such a way that it is reserved for what we call man, what is there to say? But if one reinscribes language in a network of possibilities that do not merely encompass it but mark it irreducibly from the inside, everything changes. I am thinking in particular of the mark in general, of the trace, of iterability, of *différance*. These possibilities or necessities, without which there would be no language, *are themselves not only human*. It is not a question of covering up ruptures and heterogeneities. I would simply contest that they give rise to a single linear, indivisible, oppositional limit, to a binary opposition between the human and the infra-human. And what I am proposing here should allow us to take into account scientific knowledge about the complexity of "animal languages," genetic coding, all forms of marking within which so-called human language, as original as it might be, does not

allow us to "cut" once and for all where we would in general like to cut. (116–17)

But it is not simply a matter of contesting humanism's traditional notion of language—of reconceiving language itself in terms of the dynamics of *différance* that, because they are fundamentally inhuman in both their technicity and their extension to extrahuman processes of communication, institute the inhuman at the human's very origin. For once *that* stratagem of humanism has been met, there remains the privileged *relation to that relation* that more contemporary, sophisticated forms of humanism of the sort we find in Lacan and Lévinas have reserved for themselves. As Derrida explains in "The Animal That Therefore I Am," philosophers from Aristotle to Lacan, Kant, Heidegger, and Lévinas all "say the same thing: the animal is without language. Or more precisely unable to respond, to respond with a response that could be precisely and rigorously distinguished from a reaction" (48–49). To "respond" rather than merely "react," one must be capable of "erasing," and "even those who, from Descartes to Lacan, have conceded to the said animal some aptitude for signs and for communication, have always denied it the power to *respond*—to *pretend*, to *lie*, to *cover its tracks* or *erase* its own traces"—hence the fallback position of humanism (as in Lacan) that it is the difference between communication and metacommunication, signifying and signifying *about* signifying, the ability *to lie by telling the truth*, as Lacan puts it—that surely distinguishes the human from the animal. But as Derrida notes, even if we concede that this is a more compelling distinction between human and animal than simply language use as such, it is nonetheless deeply problematic in one fundamental sense: "The fact that a trace can always be erased, and forever, in no way means—and this is a critical difference—that someone, man *or* animal, *can of his own accord* erase his traces" (50).

That point is amplified in Derrida's recent essay "And Say the Animal Responded?" which discloses just how deeply embedded Lacan is in the Cartesian tradition's conceptualization of the animal as something that can only "react" and not "respond."[32] As Derrida notes there, if we take Lacan's concept of the unconscious at its word, "the logic of the unconscious is founded on a logic of repetition which, in my opinion, will always inscribe a destiny of iterability, hence some automaticity of the reaction in every response" (202), hence eroding "the purity, the rigor, and the indivisibility of the frontier that separates—already with respect to 'us-humans'—reaction from response" (200). Moreover, if we take seriously Lacan's formulation of "the subject of the signifier," then what is disclosed in Lacan's insistence that "pretending to pretend" distinguishes the human from the animal is a residual but powerful desire in Lacan (as in humanism generally) to conceive the subject as in some sense free from the play of the trace structure of signification. In this light, as Derrida puts it, "to be subject of the signifier is also to be a subjecting subject, a subject as *master*, an active and deciding subject of the signifier, having in any case sufficient mastery to be capable of pretending to pretend and hence of being able to put into effect one's power to destroy the trace" (208). All of which, Derrida writes, "is why so long ago I substituted the concept of trace for that of signifier" (214); and all of which makes it clear that in Lacan "every reference to the capacity to erase the trace still speaks the language of the conscious, even imaginary ego" (216).

The specific moment in Derrida's intervention is crucial, I think, for a couple of reasons. First, it helps to make clear how it is that Derrida is interested in the historical and institutional specificity—not "merely," as it were, the ontological problematics—of the question of the animal. Here Beardsworth's objection in *Derrida and the Political* about Derrida's ethical formalism is worth revisiting in light of Derrida's later work on the animal and the "trace beyond the human." Beardsworth calls on Derrida to engage more directly the question of the trace and technicity as it relates to contemporary technoscience, since the latter constitutes an unprecedented *speeding up* of the dynamic relationship between the human and the technical that "risks reducing the *différance* of time, or the aporia of time"—whose very excess constitutes the "promise" of the impossible "we" to come to which any form of political organization is ethically responsible (146)—"to an experience of time that *forgets* time" (148). But what we find in Derrida's later work—and above all for Beardsworth in *Of Spirit*—is an underestimation of "the speed with which the human is losing its experience *of* time," with the result that the "promise" of ethics and politics ends up *appearing too formal, freezing Derrida's deconstructions . . . which turn the relation between the human and the technical into a 'logic' of supplementarity without history (the technical determinations of temporalization)" (154). Thus for Beardsworth "there are, consequently, 'two' instances of 'radical alterity' here which need articulation, and whose relation demands to be developed: the radical alterity of the promise and the radical alterity of the other prior to the ego of which one modality (and increasingly so in the coming years) is the technical other" (155).

But *only one modality*, I would hasten to add. Indeed, it seems likely to me (though there is no way, strictly speaking, to *argue* the point) that Beardsworth's call for "the promise to appear *through* the relation between the human and the nonhuman" (156) gets rerouted in much of Derrida's later work—*especially* in *Of Spirit*—via the question of the animal. Beardsworth asks, "With attention to the radical alterity of time, do Derrida's earlier analyses of originary technicity become eclipsed? If not . . . then *how* does one develop the relations between the promise and originary technicity?" (153). The answer, it seems to me, is via the question of the animal, *precisely* with the intention of developing a concept of the promise that is not once again automatically exclusive of nonhuman others. For Derrida would surely ask Beardsworth whether *his* concept of the radical alterity of time in this instance is not symptomatic of the humanism with which Derrida takes issue in "The Animal That Therefore I Am," in his meditations on the shared passivity, anguish, and vulnerability of the human *and* the animal in relation to death. In his later work Derrida's strategy, I suggest, is exactly the reverse of what Beardsworth calls for: attention to the question of the qualitative transformation of time by way of attention not to the speed of technoscience, but to what we might think of as the "slowness" of the animal other.

Here time, rather than being "for" the human—*even in* the form of its inhumanity in technicity, to which the human nevertheless maintains a privileged relation—instead consists of a radical asynchronicity. This is so horizontally, in evolutionary qualities and tendencies that persist across species lines (the facts of our mammalian being, of "involuntary" physiological traits and gestural repertoires, the experience of disease and, most important, the death that fatefully links the world of human and animal), and vertically, in the differences between species in the power over time—their ability to compress time, if you will, for adaptive advantage—available differentially in species-specific technicities (including, of course, the technicity of the body as the first tool, but also of the brain and the tool proper, with its apotheosis in technoscience).

In these terms, one might think of the speed and compression of time that Beardsworth (following Bernard Stiegler) associates with the historically, humanly specific phenomenon of technoscience as part of a larger evolutionary process of chronicities and periodicities in which all animals participate in a sort of shared passivity of scarcity in the face of time's alterity. For example, as J. T. Fraser has argued in *Of Time, Passion, and Knowledge*, all animals strive to increase their control over ever longer periods of future time in the interest of anticipating and adapting to changes in their environment. The differences between species may thus be described in terms of the ability to handle increased temporal complexity and the constant introduction of novel periodicities into the environment as organisms constantly adjust to each others' increasingly well honed periodicities by

introducing ever more efficient ones of their own, leading to a supersatura-
tion of chronicities that in turn generates a scarcity of *time* that drives the
evolutionary process.[33]

From this vantage—to return to the relation between time and technicity—what Derrida's work on the animal would stress is the *inhuman* rather
than the *human* relation *to the inhumanity of time and technicity itself.* This is
what Derrida means, I believe—in a formulation germane to Beardsworth's
own historicism—when as writes that "as for history, historicity, even his-
toricality, those motifs belong precisely . . . to *this* auto-definition, *this* auto-
apprehension, *this* auto-situation of man or of the human *Dasein* with respect to what is living and with respect to animal life; they belong to this
auto-biography of man that I wish to call into question today" (37).

This does not mean, however, that Derrida is not attuned to the historical specificity of our relation to animals. Indeed, "The Animal That Therefore I Am" is even more striking than "'Eating Well'" in the forthrightness
with which it meets this question. There he argues that "for about two centuries" we have been involved at "an alarming rate of acceleration" in a
transformation of our experience of animals (36), in which our

> traditional forms of treatment of the animal have been turned upside
> down by the joint developments of zoological, ethological, biological
> and genetic *forms of knowledge* and the always inseparable *techniques* of
> intervention . . . by means of farming and regimentalization at a de-
> mographic level unknown in the past, by means of genetic experimen-
> tation, the industrialization of what can be called the production for
> consumption of animal meat, artificial insemination on a massive scale,
> more and more audacious manipulations of the genome, the reduction
> of the animal not only to production and overactive reproduction (hor-
> mones, crossbreeding, cloning, etc.) of meat for consumption but also
> of all sorts of other end products, and all that in the service of a certain
> being and the so-called human well-being of man. (38)

For Derrida, "no one can seriously deny the disavowal that this involves . . .
in order to organize on a global scale the forgetting or misunderstanding of
this violence that some would compare to the worst cases of genocide" (39).
But this genocide takes on a particular, historically specific form. As Derrida puts it in one of the more striking passages in all of his work on animals,

> It is occurring through the organization and exploitation of an arti-
> ficial, infernal, virtually interminable survival, in conditions that pre-
> vious generations would have judged monstrous, outside of every

supposed norm of a life proper to animals that are thus exterminated by means of their continued existence or even their overpopulation. As if, for example, instead of throwing people into ovens or gas chambers, (let's say Nazi) doctors and geneticists had decided to organize the overproduction and overgeneration of Jews, gypsies and homosexuals by means of artificial insemination, so that, being more numerous and better fed, they could be destined in always increasing numbers for the same hell, that of the imposition of genetic experimentation, or exter-mination by gas or by fire. In the same abattoirs. (39)

It is in response to this historically specific transformation of our relations with animals that "voices are raised—minority, weak, marginal voices, little assured of their discourse, of their right to discourse and of the enactment of their discourse within the law, as a declaration of rights—in order to protest, in order to appeal . . . to what is still presented in such a problem-atic way as *animal rights*." Indeed, the value of animal rights, however prob-lematic its formulation may be, is that it calls on us to recognize how this transformation, this cruelty and disavowal on an unprecedented scale, "in-volves a new experience of this compassion," has opened anew "the immense question of pathos," of "suffering, pity and compassion," and "the place that has to be accorded to the interpretation of this compassion, to the sharing of this suffering among the living, to the law, ethics, and politics that must be brought to bear upon this experience of compassion" (40).

DISARTICULATING LANGUAGE AND SPECIES: MATURANA AND VARELA (AND BATESON)

One advantage of Derrida's formulation of the "trace beyond the human" is that it allows us not only to "move from the 'ends of man,' that is the con-fines of man, to the 'crossing of borders' between man and animal" ("The Animal That Therefore I Am," 4), but also to make an interdisciplinary crossing between philosophy and the sciences. As Eva Knodt has pointed out, the exploration of the possible convergences between the "two cul-tures" of science and the humanities "remains blocked as long as difference is modeled upon linguistic difference, and linguistic self-referentiality is considered the paradigm for self-referentiality generally." Here, of course, a good deal depends on how one understands Derrida's notions of writing and textuality. But in any case we would need to distinguish, I think, between what Knodt calls the "pan-textualist assumptions" of Derrida's formulations and those of Lyotard—not just on the question of language, but also on the question of science.[34] For Derrida's theorization of language in terms of the inhuman trace pushes in a fundamental sense in exactly the opposite direction from Lyotard's strongly vertical sense of language and seems in many ways closer to more sophisticated contemporary notions of *communication* as an essentially ahuman dynamic. Here one would eventually want to distinguish between second-wave systems theory of the sort we find in Niklas Luhmann or Humberto Maturana and Francisco Varela—for whom difference is "not 'noise' that occludes the brighter pattern to be captured in its true essence" or "a step toward something else" but is rather "how we arrive and where we stay"—and earlier theories with which Derrida, we can be sure, would have little patience.[35] In any event, it is worth lingering over the point for a moment, because Lyotard seems to prevent himself from radicalizing his concept of language in this direction precisely because of his suspicion (in *The Postmodern Condition*) of the sciences and, especially, of systems theory—the very domain of contemporary science in which the models of communication and meaning closest to those of poststructuralism have been developed.

Here, then, my aim is to give some substance to Derrida's own very general suggestions that such disciplinary crossings be pursued, as he reminds us when he protests Heidegger's dogmatic humanism toward the animal in the face of a growing and highly differentiated "zoological knowledge" ("*Geschlecht* II," 173). In doing so, I am not so much taking issue with Derrida as taking him at his word. For as he has contended episodically but steadily throughout his writings, from the early work on technicity of the late sixties and early seventies to the more recent investigations of the question of the animal, "the word *trace* must refer to itself a certain number of contemporary discourses whose force I intend to take into account," and not just in philosophy, but "in all scientific fields, notably biology, this notion seems currently to be dominant and irreducible.[36] Yet when we move the discussion into this register of the linguistic behaviors of (at least some) animals, we need to remind ourselves, as Derrida is quick to point out, that it is not simply a question of "giving language back to the animal." Rather, it entails showing how the difference in *kind* between human and animal that humanism constitutes on the site of language may instead be thought of as difference in *degree* on a continuum of signifying processes disseminated in a field of materiality, technicity, and contingency, of which "human" "language" is but a specific, albeit highly refined instance.

As he puts it in the recent essay on Lacan and the animal, "it is difficult, as Lacan does, to reserve the differentiality of signs for human language only, as opposed to animal coding. What he attributes to signs that, 'in a language' understood as belonging to the human order, 'take on their value from their relations to each other' and so on, and not just from the 'fixed correlation' between signs and reality, can and must be accorded to any code, animal or human" ("And Say the Animal Responded?" 198). And when we recall Derrida's contention that "the structure of the trace is such that it cannot be in anyone's power to erase it and especially not to 'judge' its erasure," we may now understand the full force of Derrida's argument that the trace structure of signification crosses species boundaries and exceeds the question of the subject, human or animal. Derrida's deconstruction of the ability to "erase" one's traces "might appear subtle and fragile but its fragility renders fragile all the solid oppositions that we are in the process of tracking down [dé-pister], beginning with that between symbolic and imaginary which underwrites finally this whole anthropocentric reinstitution of the superiority of the human order over the animal order" ("And Say the Animal Responded?" 216–17). In other words, to recall Derrida's admonition, "The animal, what a word!" is to remember that while the question of signifying behaviors may seem relevant only for some animals in particular— namely those, such as the great apes, in whom linguistic behaviors have been observed—the larger point is that this reopening of the question of language has enormous implications for the *category* of the animal in general— the animal in the "singular," as Derrida puts it—and how it has traditionally been hypostatized over and against the human—again in the singular.

But if my aim is to use work in contemporary science to put some meat on the bones of Derrida's rather general observations on the subjection of both human and animal to the force of the trace, I also want to bring to bear on work in contemporary science—here, the work of Maturana and Varela—the force of the extraordinarily searching investigations of questions of ethics that we find in contemporary philosophy. For as I have argued elsewhere,[37] Maturana and Varela's own rendering of the ethical implications of their work in biology and epistemology seems to smuggle back in a very traditional form of humanism that their work in epistemology and biology promises to move us beyond.

With these qualifications in mind, I want to turn aside and consider as briefly as I can the work of Maturana and Varela on animals, language, and what they call the emergence of "linguistic domains." I have no intention, of course, of surveying what has become the immense field within ethology of animal language studies.[38] And though I will turn very briefly to these issues at the end of this chapter, I will largely be ignoring very complex questions of institutional disciplinarity in the relations between science and philosophy, questions that would no doubt require their own extended and very different investigation. Similarly, I will be postponing until another occasion a detailed comparison of the theories of meaning in poststructuralism and contemporary systems theory (the latter has received its most sophisticated elaboration in the work not of Maturana and Varela but of Niklas Luhmann).

For now, however, I want to examine the theoretical frame Maturana and Varela have provided for understanding the relations of animals, humans, and language. For them, the baseline physiological structure that an animal must possess to provide the physical basis for the emergence of "third-order structural couplings" and, within that, "linguistic domains" is sufficient cephalization—that is, a certain concentration and density of neural tissue. As they put it, "The function of the nervous system diversifies tremendously with an increase in the variety of neuronal interactions, which entails growth in the cephalic portion. . . . In other words, this increase in cephalic mass carries with it enormous possibilities for structural plasticity of the organism. This is fundamental for the capacity to learn."[39] For Maturana and Varela, learning and what we usually call "experience" are precisely the result of "structural changes" within the nervous system, and specifically within the synapses and their "local characteristics" (167). Unlike mechanical cybernetic systems, even those that are capable of elementary forms of reflexivity and self-monitoring (artificial intelligence systems, for example), biological systems are self-developing forms that creatively reproduce themselves by embodying the processes of adaptive changes that allow the organism to maintain its own autonomy or "operational closure." For Maturana and Varela—and this is the theoretical innovation for which they are best known—all living organisms are therefore "autopoietic" unities; that is, they are "continually self-producing" according to their own internal rules and requirements, which means they are in a crucial sense *closed* and self-referential in terms of what constitutes their *specific* mode of existence, even as they are *open* to the environment on the level of their material structure. As they explain it,

Autopoietic unities specify biological phenomenology as the phenomenology proper to those unities with features distinct from physical phenomenology. This is so, not because autopoietic unities go against any aspect of physical phenomenology—since their molecular components must fulfill all physical laws—but because the phenomena they generate in functioning as autopoietic unities depends on their organization and the way this organization comes about, and not on the physical nature of their components (which only determine their space of existence). (*Tree*, 51)

The nervous system, for example,

> does not operate according to either of the two extremes: it is neither representational nor solipsistic. It is not solipsistic, because as part of the nervous system's organism, it participates in the interactions of the nervous system with its environment. These interactions continuously trigger in it the structural changes that modulate its dynamics of states. . . . Nor is it representational, for in each interaction it is the nervous system's structural state that specifies what perturbations are possible and what changes trigger them. (169)

This is the view widely held in neurobiology and cognitive science, where most scholars now agree—to take perhaps the most often-cited example, color vision—that "our world of colored objects is literally independent of the wavelength composition of the light coming from any scene we look at. . . . Rather, we must concentrate on understanding that the experience of a color corresponds to a specific pattern of states of activity in the nervous system which its structure determines" (21–22). For Maturana and Varela, then, the environment does not present stimuli to the organism, replete with specifications and directions for appropriate response in an input-output model. As they describe it, "the structure of the environment only *triggers* structural changes in the autopoietic unities (it does not specify or direct them)" (75). "In this way," they continue, "we refer to the fact that the changes that result from the interaction between the living being and its environment are brought about by the disturbing agent but *determined by the structure of the disturbed system*" (96). This means that "the nervous system does not 'pick up information' from the environment, as we often hear. On the contrary, it brings forth a world by specifying what patterns of the environment are perturbations and what changes trigger them in the organism" (169). It is this break with the representational model that distinguishes the work of Maturana and Varela from most of even the most sophisticated work on self-organizing systems in the sciences—a fact whose full epistemological implications I will touch on below.

In animals with sufficient cephalization and plasticity, it is possible for "interactions *between* organisms to acquire in the course of their ontogeny a *recurrent* nature" (180), and only with reference to that specific ontogeny, in its various degrees of contingency and uniqueness, can we understand the behavior of such animals. When these interactions between specific ontogenies become recurrent, organisms develop a *"new phenomenological domain"* (180): *"third-order structural couplings"* (181), or "social life for short" (189). As Maturana and Varela put it, what is common to third-order unities is that "whenever they arise—if only to last for a short time—they generate a particular internal phenomenology, namely, one in which *the individual ontogenies of all the participating organisms occur fundamentally as part of the network of co-ontogenies that they bring about in constituting third-order unities*" (193).[40]

In these instances, the evolutionary problem immediately becomes how, given such variation, the social animal will maintain the autopoiesis of the social structure. The answer, in a word, is *communication* (196, 198–99)—and communication in the specific antirepresentationalist sense I have already touched on. To understand the relation between the broader phenomenon of communication and the more specific matter of language as such, it might be useful to contrast the communication of relatively nonplastic social animals, the social insects, with those of more plastic animals such as wolves or humans. In the case of the insects, communication can take place by a few direct chemical signals (trophallaxis), because the behavior to be regulated is not susceptible to great ontogenic variation. When the reverse is true, however—when ontogenic variation must be not just tolerated but in fact made productive for the autopoiesis of the social structure—then the animal must learn "acquired communicative behaviors" that depend on its individual ontogeny. When this happens, the animal is engaged in the production of a "linguistic domain," behaviors that "constitute the basis for language, but . . . are not yet identical with it" (207).[41]

Even though human beings are not the only animals that generate linguistic domains, "what is peculiar to them is that, in their linguistic coordination of actions, they give rise to a new phenomenal domain, viz. the *domain of language*. . . . In the flow of recurrent social interactions, language appears when the operations in a linguistic domain result in coordinations of actions about actions that pertain to the linguistic domain *itself*" (209–10). "In other words," they conclude, "we are in language or, better, we 'language,' only when through a reflexive action we make a linguistic distinction of a linguistic distinction" (210). Now this view of the specificity of language as *metalinguistic*—as the ability to make linguistic distinctions about linguistic distinctions that separates the human from the animal—may at first glance seem similar to some of the familiar strategies of humanism that we have examined in the foregoing pages (the Lacanian view critiqued by Derrida, for example). Here, however, Maturana and Varela emphasize that the relation between linguistic domains, the emergence of language per se, and species is dynamic and fluid, one of degree and not of kind. It is not an ontological distinction, in other words, even if it is a phenomenological one. As they are quick to point out, "cogent evidence" now shows that other animals (most famously, great apes) are "capable of interacting with us in rich and even recursive linguistic domains" (212). More than that, it seems that in many of these instances animals are indeed capable of "making linguistic distinctions of linguistic distinctions"—that is, of languaging.[42] For them, "language is "a permanent biologic possibility in the natural drift of living beings" (212).

The point here, of course, is not to determine whether animals can "make all the linguistic distinctions that we human beings make" (215), but to rigorously theorize the *disarticulation* between the category of language and the category of species, for only if we do so can the relation between human, animal, and language be theorized in *both* its similarity and its difference. For example, drawing on language experiments with chimps, they argue that animals equipped with a signifying repertoire, like humans, develop their ability to participate in linguistic domains in proportion to their interpersonal interactions with other languaging beings (217). When they are permitted to live in an environment rich in opportunities for "linguistic coupling," they can communicate and express their subjectivities in ways more and more identifiably like our own—which suggests, of course, that such subjectivities are not given as ontological differences in kind but emerge as overlapping possibilities and shared repertoires in the dynamic and recursive processes of their production. (And the reverse is true as well; when animals and humans are deprived of opportunities for third-order couplings in social interactions and communications, their behaviors become more mechanical and "instinctive," as their ontogenies are severely limited and invariable.)[43]

We can gain an even finer-grained sense of how systems theory thinks of this relation by turning briefly to the work of Gregory Bateson. As he points out in his analysis of "play" among mammals, this phenomenon "could only occur if the participant organisms were capable of some degree of metacommunication, *i.e.*, of exchanging signals which would carry the message 'this is play.'"[44] "The playful nip denotes the bite," he continues, "but it does not denote what would be denoted by the bite"—namely aggression or fight (181). What we find here, as in other behaviors among animals such as "threat," "histrionic behavior," and "deceit," is what Bateson calls "the primitive occurrence of map-territory differentiation," which "may have been an important step in the evolution of communication." As he explains,

> Language bears to the objects which it denotes a relationship comparable to that which a map bears to a territory. Denotative communication as it occurs at the human level is only possible *after* the

evolution of a complex set of metalinguistic (but not verbalized) rules which govern how words and sentences shall be related to objects and events. It is therefore appropriate to look for the evolution of such metalinguistic and/or meta-communicative rules at a prehuman and preverbal level. (180)

As Bateson points out, however, it is not as if such instances are simply transcended by the advent of specifically human modes of verbal interaction, for "such combinations as histrionic play, bluff, playful threat" and so on "form together a single total complex of phenomena" that we find not only in various childhood patterns of behavior, but also in adult forms such as gambling, risk taking, spectatorship, initiation and hazing, and a broad range of ritualistic activities—all examples of "a more complex form of play: the game which is constructed not upon the premise 'This is play' but rather around the question 'Is this play?'" In all of these we find more elaborate forms of the map-territory relation at work in mammalian play generally, where "paradox is doubly present in the signals which are exchanged. . . . Not only do the playing animals not quite mean what they are saying but, also, they are usually communicating about something which does not exist" (182). The playful baring of the fangs between two wolves, for example, signifies the bite that does not exist; but the bite that does not exist itself signifies a *relationship*—in this case of dominance or subordination—whose "referent," if you will, is itself the autopoiesis of the pack structure that determines those relationships.

Indeed, as Bateson argues, mammalian communication in general is "primarily about the rules and the contingencies of relationship." For example, the familiar movements a cat makes in "asking" you for food are, behaviorally speaking, essentially those that a kitten makes toward a mother cat. And "If we were to translate the cat's message into words, it would not be correct to say that she is crying 'Milk!' Rather, she is saying something like 'Mama!' Or perhaps, still more correctly, we should say that she is asserting 'Dependency! Dependency!'" From here, "it is up to you to take a *deductive* step, guessing that it is milk that the cat wants. It is the necessity for this deductive step" (and this strikes me as a singularly brilliant insight) "which marks the difference between preverbal mammalian communication and *both* the communication of bees and the languages of men" (367).

For Bateson, it may be that "the great new thing" in the evolution of human language is not "the discovery of abstraction or generalization, but the discovery of how to be specific about something other than relationship"— to be denotative about actions and objects, for example. But what is equally remarkable is how tied to the communication of preverbal mammals human communication continues to be (367). Unlike the digital mode of communication typical of verbal languages, in which the formal features of signs are not driven "from behind" by the real magnitudes they signify ("The word 'big' is not bigger than the word 'little,'" to use Bateson's example), in the analogical form of kinesic and paralinguistic communication used by preverbal mammals "the magnitude of the gesture, the loudness of the voice, the length of the pause, the tension of the muscle, and so forth commonly correspond (directly or inversely) to magnitudes in the relationship that is the subject of discourse" (374), and they are signaled via "bodily movements," "involuntary tensions of voluntary muscles," "irregularities of respiration," and the like. "If you want to know what the bark of a dog 'means,' you look at his lips, the hair on the back of his neck, his tail, and so on" (370).

It may be, as Bateson argues, that human languages have a few words for relationship functions, "words like 'love,' 'respect,' 'dependency,'" but these words function poorly in the actual discussion of relationship between participants in the relationship. "If you say to a girl, 'I love you,' she is likely to pay more attention to the accompanying kinesics and paralinguistics than to the words themselves" (374). In other words—and here I think we should remember Cavell's discussion of "skeptical terror"—she will look for the involuntary message your body is sending in spite of you, since "discourse about relationship is commonly accompanied by a mass of semivoluntary kinesic and autonomic signals which provide a more trustworthy comment on the verbal message" (137). This is why, according to Bateson, we "have many taboos on observing one another's kinesics, because too much information can be got that way" (378). And, one might add by way of an example many of us have experienced, it is also the very absence of these cues that makes e-mail such an unnerving and explosive form of communicative exchange—there is no damping or comparative modulation of the digital message by any accompanying analogical signals.

Bateson's work on language, communication, and species helps amplify and elaborate what Derrida has in mind, I think, in his formulation of the trace beyond the human, and this in two senses: first, in evolutionary terms, as the outcome of processes and dynamics not specifically or even particularly human that remain sedimented and at work in the domain of human language broadly conceived; and second, in terms of how language is traced by the material contingency of its enunciation in and through the body, in its "involuntary" kinesic and paralinguistic significations that speak in and through us in ways that the humanist subject of "intention" and "reflection"

cannot master, ways that link us to a larger repertoire and history of signification not specifically human and yet intimately so.

This view of language has important implications for our ability to theorize the continuities between human and animal subjectivities in relation
to the emergence of linguistic domains, while respecting the differences.
Bateson argues that "the discrimination between 'play' and 'nonplay,' like
the discrimination between fantasy and nonfantasy, is certainly a function
of secondary process, or 'ego.'" The ability to distinguish between play and
nonplay—the ability to make statements whose paradoxical status of the
sort we find in play is a direct result of an organism's understanding and ma-
nipulation of a metacommunicative frame—is directly related to the emergence of something like subjectivity (Bateson's "ego") as a dynamic that is
recursively tied to the evolution of increasingly complex communicative behaviors (185). For example, Maturana and Varela discuss a well-known experiment in which a gorilla is shown his reflection in a mirror, is anes-
thetized and has a red dot painted between his eyes, and is then awakened
and shown his reflection again, at which point the ape immediately, on seeing the dot, points to his own forehead—not that of the mirror image. "This
experiment," they argue, "suggests that the gorilla can generate a domain of
self through social distinctions. . . . How this happened we do not know. But
we presume it has to do with conditions similar to those leading to the evolution of human linguistic domains" (224–25). "It is in language," they continue,

> that the self, the I, arises as the social singularity defined by the oper-
> ational intersection in the human body of the recursive linguistic dis-
> tinctions in which it is distinguished. This tells us that in the network
> of linguistic interactions in which we move, *we maintain an ongoing de-
> scriptive recursion which we call the "I." It enables us to conserve our lin-
> guistic operational coherence and our adaptation in the domain of language.*
> (231)[45]

This processive, recursive, antirepresentational account of the relation
between material technicities, linguistic domains, and the emergence of
subjectivities has the advantage of allowing us to address the specificity of
our similarities and differences with other creatures—especially those creatures who are enough like us to complicate and challenge our discourses of
subjectivity—but without getting caught in the blind alleys of "intention"
or "consciousness" (or what amounts to the same thing on methodological
terrain in the sciences, "anthropomorphism") that have plagued attempts to
understand in what specific sense we share a world with nonhuman animals.

All of this is summed up nicely, I think, by philosopher and cognitive scientist Daniel Dennett when he writes that language "plays an enormous role in the structuring of a human mind, and the mind of a creature lacking language—and having really no need for language—should not be supposed to be structured in these ways. Does this mean that languageless creatures 'are not conscious at all' (as Descartes insisted)?" No, because to put the question that way presupposes

> the assumption that consciousness is a special all-or-nothing property that sunders the universe into vastly different categories: the things that have it . . . and the things that lack it. Even in our own case, we cannot draw the line separating our conscious mental states from our unconscious mental states. . . . [W]hile the presence of language marks a particularly dramatic increase in imaginative range, versatility, and self-control . . . these powers do not have the *further* power of turning on some special inner light that would otherwise be off. (447)

This does not mean that language is not ethically to the point. Quite the contrary. Indeed, it is worth articulating the specificity of this linkage if only because a persistent problem in contemporary theory has been theorizing the *specificity or singularity* of *particular* animals and the ethical issues that attend those specific differences. In contemporary theory—I am thinking here especially of the important work by Gilles Deleuze and Félix Guattari that I will take up later—the power and importance of the animal is almost always its pull toward a *multiplicity* that operates to unseat the singularities and essentialisms of identity that were proper to the subject of humanism. But this is of little help in addressing the ethical differences between abusing a dog and abusing a scallop—differences that seem, to many people, to be to the point, even if they are certainly not ethically the *only* point (in which case considerations of biodiversity and the like might come into play as well).

Revisiting, as we saw Derrida do earlier, Jeremy Bentham's critique of Descartes—the question is not can animals talk, or can they reason, but can they *suffer*—Dennett argues that while languaging and suffering "usually appear to be opposing benchmarks of moral standing," in fact it make sense to argue that the greater an animal's capacities in the former regard, the greater its capacities in the latter, "since the capacity to suffer is a function of the capacity to have articulated, wide-ranging, highly discriminative desires, expectations, and other sophisticated mental states" (449)—which helps explain the intuitive sense most of us have that the suffering of a horse or a dog is a weightier matter than that of a crayfish. "The greater the scope, the richer the detail, the more finely discriminative the desires, the worse it is when those desires are thwarted," he continues. "In compensation for having to endure all the suffering, the smart creatures get to have all the fun. You have to have a cognitive economy with a budget for exploration and self-stimulation to provide the space for the recursive stacks of derived desires that make fun possible. You have taken a first step"—and here we should recall Maturana and Varela's "linguistic distinction of a linguistic distinction"—"when your architecture permits to you to appreciate the meaning of 'Stop it, I love it!' Shallow versions of this building power are manifest in some higher species, but it takes a luxuriant imagination, and leisure time—something most species cannot afford—to grow a broad spectrum of pleasures" (450).

And yet Dennett, like Bateson, remains tied to an essentially representationalist frame, one that continues to believe that the question of an "objective" or "correct" interpretation of heterophenomenological worlds is essentially unproblematic. Aside from the epistemological problems that such a position has on its own terms—problems I have discussed elsewhere in some detail[46]—it is only when that frame is rigorously dismantled, I believe, that there can be fruitful interdisciplinary interchange of the sort we can generate between Derrida and Maturana and Varela. Indeed, as I want to argue now, to believe that organisms internalize the environment in the form of "representations" or even "information" is to have already committed the kind of Cartesian hubris diagnosed by Derrida in "The Animal That Therefore I Am," because this putatively "objective" or "realist" view of the world—the world of which organisms have more or less "accurate" representations depending on the sophistication of their filtering mechanisms—is, despite appearances, referenced to an idealism founded on the fantasy that human language is sovereign in its mastery of the multiplicity and contingency of the world—the fantasy, to put it in the hybrid terms I am using here, that there is such a thing as non(self)deconstructible observation.

More specifically, to return to Maturana and Varela, the nervous system may operate by way of its own autopoietic closure, but "we as observers have access both to the nervous system and to the structure of its environment. We can thus describe the behavior of an organism as though it arose from the operation of its nervous system with representations of the environment or as an expression of some goal-oriented process. These descriptions, however, do not reflect the operation of the nervous system itself. They are good only for the purposes of communication among ourselves as observers" (*Tree*, 132). To say as much confronts us with "a formidable snag," however, because "it seems that the only alternative to a view of the nervous system as operating with representations is to deny the surrounding reality" (133). The way out of this dilemma, they contend, is to confront it head-on in the distinction between what Niklas Luhmann calls first-order and second-order observation. In first-order observation, we are dealing with the observation of objects and events—a territory, to use Bateson's metaphor—in terms of a given map or code based on a fundamental, constitutive distinction that organizes the code. In second-order observation we are observing observations—and observing, moreover, how those observations are constructed atop a blindness to the wholly contingent nature of their constitutive distinction. (The legal system, for example, cannot carry out its observations of legal versus illegal while at the same time recognizing the essential identity of both sides of the distinction, its essential tautology: legal is legal.) As Dietrich Schwanitz puts it, first-order observation is "unable to observe the distinction on which it bases its own observation. This is observation's 'blind spot.' Therefore, observation is also unable to see whatever has been excluded from observation by its distinction. If observation is to be made observable, it is necessary to bring about a change of distinction, a displacement of the difference—in other words, a kind of deconstruction."[47]

"As observers," Maturana and Varela explain,

> we can see a unity in *different* domains, depending on the distinctions we make. Thus, on the one hand, we can consider a system in that domain where its components operate, in the domain of its internal states and structural changes. . . . On the other hand, we can consider a unity that also interacts with its environment and describe its history of interactions with it. . . . Neither of these two possible descriptions is a problem per se: both are necessary to complete our understanding of a unity. It is the observer who correlates them from his outside perspective. . . . The problem begins when we unknowingly go from one realm to another and demand that the correspondences we establish between them (because we see these two realms simultaneously) be in fact a part of the operation of the unity. (135–36)

If this sounds circular, it is. But it is precisely that circularity that provides the bridge between the second-order systems theory of Maturana and Varela and the deconstruction of Derrida. And it is also this very circularity that prevents the relation between physical substratum (cephalizaton) and phenomenological domain (languaging) in Maturana and Varela from devolving into a type of positivism.

Acknowledging the "slightly dizzy sensation" that attends "the circularity entailed in using the instrument of analysis to analyze the instrument of analysis," Maturana and Varela write that "*every act of knowing brings forth a world*" because of the "inseparability between a particular way of being and how the world appears to us." For us, as languaging beings, this means that "every reflection, including one on the foundation of human knowledge, invariably takes place in language, which is our distinctive way of being human and being humanly active" (26). Or as Maturana puts it elsewhere in an especially exacting formulation:

> Contrary to a common implicit or explicit belief, scientific explanations . . . constitutively do not and cannot operate as phenomenic reductions or give rise to them. This nonreductionist relation between the phenomenon to be explained and the mechanism that generates it is operationally the case because the actual result of a process, and the operations in the process that give rise to it in a generative relation, *intrinsically take place in independent and nonintersecting phenomenal domains.* This situation is the reverse of reductionism. . . . [This] permits us to see, particularly in the domain of biology, that there are phenomena like language, mind, or consciousness that require an interplay of bodies as a generative structure but do not take place in any of them. In this sense, science and the understanding of science lead us away from transcendental dualism.[48]

What Maturana and Varela offer here, I think, is their own version of how, as in Derrida's account (to borrow Rodolphe Gasché's characterization), the conditions of possibility for discourse are at the same time conditions of impossibility.[49] More precisely, we can insist on the "independent and nonintersecting phenomenal domains" that make up existence and that thus, in *being nonintersecting*, defy the mastery of any concept, identity, or *logos*, but we can do so only by means of the phenomenal domain of language. As Eva Knodt has pointed out, for both systems theory and deconstruction, there is no "beyond" of language, "and the proposal to move from a linguistic to a systems-theoretical paradigm should not be construed as an attempt to escape the problem of linguistic self-referentiality" (xxxii).

For Maturana and Varela, however—and this, I think, captures the full force of Derrida's radicalizing of the concept of the "trace beyond the human" for the present discussion and perhaps marks an affinity of both over and against Luhmann—that phenomenal domain requires "an interplay of bodies as a generative structure" but does not take place in any one of them. As Maturana puts it in a formulation that, in light of Bateson's work on mammalian communication, has particular resonance for Derrida's insistence on the fundamentally ahuman character of language, on its erosion by its other, by *all its others:* "As we human beings exist in language, our bodyhood is the system of nodes of operational intersection of all the operational coherences that we bring forth as observers in our explanation of our operation" ("Science and Daily Life," 49). Hence "the bodyhood of those in language changes according to the flow of their languaging, and the flow of their languaging changes contingently to the changes of their bodyhood. Due to this recursive braiding of bodyhood changes and consensual coordinations of actions in language, everything that the observer does as a human being takes place at the level of his or her operational realization in his or her bodyhood in one and the same domain," even though different cognitive domains, such as the "practical" and the "theoretical," may "in the conversational domains in which they are distinguished as human activities" appear to be totally different (45).

Circularity in Maturana and Varela, then, leads us back to the contingency of the observer, and in two specific senses: first, an observer whose observations are constituted by the domain of language, but a domain of language that is not foundational because it is "only" the result of broader evolutionary processes not specifically linguistic at all; and second, an observer who, "recursively braided" to its bodyhood, is always already internally other and, in terms of the carnophallogocentric tradition, animal. But whereas Derrida's emphasis on the deconstructability of the observer's observation would fall on the paradoxical relation between *logos* and the internal differential dynamics of language, for Maturana and Varela the emphasis would fall instead on the paradoxical relation between the observer's self-reference and its biological heteroreference: vertically in the bodyhood of the observer and horizontally in the observer's evolutionary emergence via inhuman dynamics and mechanisms. This yields the paradoxical result that only beings like this could have emerged to provide an explanation of how beings like this could have emerged, and so on. For both, the hypostatized relation between "inside" and "outside" is thus made dynamic, a differential interplay that deontologizes as it reconstitutes. As Dietrich Schwanitz puts it, "Both theories make difference their basic category, both temporalize difference and reconstruct meaning as . . . an independent process that constitutes the subject rather than lets itself be constituted by it" (153).

In Derrida, however, the deconstructability of *logos* propels us outward toward the materiality and contingency that Maturana and Varela will associate with structure, whose demands and "triggers" constitute a problem for the autopoiesis of the organism. In this way the analyses of Derrida and of Maturana and Varela move, in a sense, in opposite directions: Derrida's from the inside out, as it were, from the originary problem of the self-reproduction of *logos* to the contingency of the trace, and Maturana and Varela's from the outside in, from the originary problem of the overwhelming contingency and complexity of the environment to the autopoiesis of self-referential organization that, by reducing complexity, makes observation possible.[50]

In would be tempting, I suppose, to find in Derrida's "trace beyond the human" the opening of a radicalized concept of language to a kind of biologization—not just "materialization," which would be Derridean enough for most Derrideans, but more pointedly, in the later work, to *"the problem of the living."* Similarly, it is tempting to find in the biology of Maturana and Varela a kind of linguisticizing of biology, in their attention to the epistemological problem that language is "our starting point, our cognitive instrument, and our sticking point" (*Tree*, 26). But here one last word from systems theory is in order; for what makes such a "convergence" possible (if one wants to put it that way) is, paradoxically, not attempting to step outside the limits of different disciplines and language games, but rather pushing them internally to their self-deconstructive conclusions. In this light, what looks at first glance like the solipsistic insistence on self-reference and operational closure in systems theory might be seen instead as in the services of what Carolyn Merchant calls a "reconstructive knowledge" based on "principles of interaction (not dominance), change and process (rather than unchanging universal principles), complexity (rather than simple assumptions)."[51] And it is in this light that we can see systems theory, as Niklas Luhmann puts it, as "the reconstruction of deconstruction."[52]

For Luhmann—to put it very schematically—we live in a "functionally differentiated" society, in which we find a horizontal proliferation of language games and social systems, none of which provides a totalizing perspective on the others, and all of which are observations that are blind to their own constitutive distinctions. The fact of this self-referential closure of language games, however, paradoxically drives them toward a kind of convergence, so that it is precisely *by* working vertically in different disciplines that Derrida and Maturana and Varela can complement one another. As Luhmann puts it in *Observations on Modernity*, what we find here is not "reciprocal impulses that could explain the expansion of certain thought dispositions," but rather an "equifinal process" "that leads to a result from different starting points and that is dissolving traditional ontological metaphysics."[53] "With all the obvious differences that result from the different functions and codings of these systems," he continues, remarkable similarities appear, such as with the "deep structures" of modern society. . . . This type of society no longer conceives of itself with preeminences of single components—with nobility or the state. The effect of the social relationship shows itself in the nonrandom consequences of the autonomy of function systems. They prove themselves to be similar despite all their differences (and in this specific sense, as modern) because they have achieved operative segregation and autonomy. This is not possible except in the form of arrangements that require, among other things, an observation of the second order [as in Maturana and Varela's separation of phenomenal domains, or Derrida's logic of the supplement] as a systems-carrying normal operation. This explains the conspicuous finding that this society accepts contingencies like none other before it. (60–61)

It may also explain how we find the *biologists* Maturana and Varela sounding a lot like the *philosopher* Derrida in *Autopoiesis and Cognition*, where they contend that

the domain of discourse is a closed domain, and it is not possible to step outside of it through discourse. Because the domain of discourse is a closed domain it is possible to make the following ontological statement: *the logic of the* description *is the logic of the* describing *(living) system (and his cognitive domain)*.

This logic demands a substratum for the occurrence of the discourse. We cannot talk about this substratum in absolute terms, however, because we would have to describe it. . . . Thus, although this substratum is required for epistemological reasons, nothing can be said about it other than what is meant in the ontological statement above.[54]

"Nothing outside the text" indeed—except, of course, everything.

PART *Two*

CHAPTER 3

Subject to Sacrifice

Ideology, Psychoanalysis, and the Discourse of Species
in Jonathan Demme's *The Silence of the Lambs*

With Jonathan Elmer

The mixture of menace and aestheticism that distinguishes Jonathan Demme's *The Silence of the Lambs* (1991) is evoked effectively by the film's publicity poster. Blossoming from Jodie Foster's mouth like an exotic *fleur du mal* is a moth (viewers of the film will recognize it as the death's-head moth that serves as a personal totem for "Buffalo Bill"), which conventionally fixes the female icon at the point where her beauty and her helplessness converge. In its position over her mouth, the moth stands for what threatens her, and it also sends us back to the film's title and its ominous key term: silence. But while the title says the silence of the *lambs*, the image conveys something more generic: the silence of the *heroine*.

Most treatments of the film have followed the redirection of attention suggested by the poster, seeing the film as part of Hollywood's confused response to shifting norms of gender and sexuality, a response that answers the unabated depiction of guns pointed at women with portrayals of women pointing guns: *The Silence of the Lambs* might thus seem to be aligned with films such as *Fatal Attraction, Blue Steel, La femme Nikita,* or *Thelma and Louise.* In this understanding, the film's ideological deep structure organizes gender and sexuality relations by giving us various mixtures for endorsement or repudiation: masculinized women (Clarice Starling [Jodie Foster]), feminized men (the aspiring transsexual "Buffalo Bill" [Ted Levine]), feminized women (the character of Catherine Martin [Brooke Smith]), even— why not?—masculinized men (Hannibal Lecter [Anthony Hopkins]).[1]

This understanding is surely correct, as far as it goes. As Carol Clover has recently demonstrated, however, horror films are quite capable of producing what might be called an ideological feint. Thus, in Clover's ingenious argument, slasher films or films of demonic possession, while apparently obsessed with the investigation and regulation of the category of the feminine, can in fact serve—through the dynamics of cross-gender identification with the "Final Girl" (the sole female survivor of the generic slasher film)—as staging grounds for unavowable forms of masculine experience—most fundamentally, masochism.[2] An important formal feature of horror fictions, then, seems to be the inducement and subsequent disguise of audience identification, and what might be called the film's manifest ideology serves to direct attention away from powerful lines of identification that thereby remain latent. It is not that the manifest ideology is not relevant: slasher films are undeniably about the category of the feminine, and *The Silence of the Lambs* cannot be understood without reference to its complicated gender discourse. Our point (and Clover's), rather, is that the horror genre, in so flagrantly eliciting ambivalent identifications, poses difficulties for the prevailing paradigms of ideology critique, which generally understand fictions to be about the reformulation and reassertion of ideological norms in the resolutions of the denouement. Such is the understanding that lies behind Fredric Jameson's influential thesis that aesthetic objects offer "fantasy bribes" to their audiences, glimpses of the utopian possibility of a less exploitative and atomized social order that draw on "the ineradicable drive towards collectivity that can be detected, no matter how faintly and feebly, in the most degraded works of mass culture," only to be all the more effectively managed at the fiction's end.[3]

But this "ineradicable drive towards collectivity," if indeed such a thing exists, can take any number of forms, some of them more dystopian than utopian. Horror fiction seems to trouble just these sorts of dichotomies, because it is not at all clear in horror whether the audience's exhilaration comes from the monster's transgression of social norms or from the reassertion of those norms by the extirpation of the monster itself. Stephen King, who should know about such things, captures this well: "Horror appeals to us because it . . . is an invitation to indulge in deviant, antisocial behavior by proxy—to commit gratuitous acts of violence, indulge our puerile dreams of power, to give in to our most craven fears. Perhaps more than anything else, the horror story says it's okay to join the mob, to become the total tribal being, to destroy the outsider."[4] The slippage of identification in King's account from monster to mob is very much to our point, for it is hard to know whether the "fantasy bribe" here is about the violation of norms or their violent retrenchment. What horror suggests for ideology critique, then, is that the ideological "point" of fictions may not lie exclusively with the reimposition of ideological norms in the fiction's ending, but rather may concern its complicated and contradictory *middle*, where identificatory energies are released and invested.[5] As we shall see, the ending of *The Silence of the Lambs* is indeed immensely important,[6] but less for what it ties together than for what it leaves hanging. More than most films, this one exhibits the way the energies aroused in the aesthetic experience of contradictory identifications are not fully recoupable by any ideological closure but rather continue, like Lecter himself, to circulate in disguise on some other scene. We want to diverge from prevailing models of ideology critique, then, by pursuing more vigorously the recognition that ideological discourses do not merely "operate" horror films but are themselves "operated on," often serving as manifest screens, or feints, that disguise latent discourses and identifications.

We retain the term "ideology" here to remind ourselves that this process always takes place within a social sphere that is "uneven" with regard not only to economic distribution and class power but also to sexuality, gender, species, and much else besides. In her later work, Judith Butler argues for the kind of attentiveness to the asymmetry between discourses that we have in view here. "It seems crucial," she writes, "to resist the model of power that would set up racism and misogyny and homophobia as parallel or analogical relations," because such a model "delays the important work of thinking through the ways in which these vectors of power require and deploy each other for the purpose of their own articulation."[7]

The unevenness of social discourses is ideologically *productive*, in other words, and it is therefore crucial, as we shall see, that in a project like *The Silence of the Lambs* the discourses of gender and class—or, in another register, psychoanalysis and aestheticism—all remain in play, powerful in their own right. Our argument, then, is that these discourses "deploy each other" in an uneven ideological space specific to this film, and that this deployment is critical to the systematic rearticulation of these codes in terms of the most far-reaching and powerful discourse in the film: the discourse of species.

To return to the film's publicity poster, it is just this sort of "deployment" that is at work as the "silence" of the film's title is realigned with Starling and directed away from the invisible lambs. For at the heart of the film's trauma, we will argue, lies not cross-gender, or cross-class, but cross-*species* identification. Initially, of course, we must understand this trauma as Starling's own, for the title refers to her unsuccessful effort to save some lambs from the slaughterhouse. "Have the lambs stopped screaming?" asks Lecter, in a tone somewhere between gentle and mocking, in his final phone call to Starling. And although this scene might seem to mark the triumph of Starling— she has saved Catherine Martin, and she has just graduated from FBI training—it is not clear that these compensatory efforts have finally overcome her trauma, for she can find no answer to Lecter's question. Here, indeed, the ambiguity of the lambs' silence merges with Starling's own. Are we to

take the lambs' silence, or Starling's silence about that silence, as the sign of
deliverance, or rather as something even more ominous than their screaming? At the end of the film Starling has "made it," acceded to the law of culture and the culture of the law. But her career triumph does not signal any
end to the regime of sacrifice imaged in the lambs' slaughter, for in the
blackly comic conclusion to the film, it is the cannibal Lecter who has the
last word: "I'm having an old friend for dinner." Here, in the face of the
film's evident endorsement of Starling's career trajectory (in which hard
work and self-sacrifice earn the reward of upward mobility), there surfaces
a sort of counterknowledge of human society as unremittingly feral, as
never-ending sacrifice, with no guaranteed payoffs. Dog Eat Dog. People
Are Sheep. Ambling at film's end down the Caribbean thoroughfare in his
bad hairpiece and baggy linen suit, Lecter becomes the postmodern wolf in
sheep's clothing.

THE SPECIES GRID AND THE LAW OF CULTURE

As we have already seen, in recent years Jacques Derrida has increasingly directed attention to what he calls the problem of the animal. Like writing and
the feminine, the animal poses grave definitional and practical threats to the
discourse of humanism, in which authority and autonomy are "attributed to
the man (*homo* and *vir*) rather than to the woman, and to the woman rather
than to the animal."[8] This hierarchy of subjects and values essentially constitutes, in Derrida's view, the "schema that dominates the concept of subject" (114). Because "carnivorous sacrifice is essential to the structure of
subjectivity," it also forms the "basis of our culture and our law" and "all the
cannibalisms, symbolic or not, that structure intersubjectivity in nursing,
love, mourning and, in truth, in all symbolic and linguistic appropriations."[9]
In order thus to "underscore the sacrificial structure" of such discourses,
Derrida has extended his famous portmanteau designation of the essence of
Western metaphysics, which he accordingly now refers to as "carnophallogocentrism": "It is a matter of discerning a place left open, in the very structure of these discourses (which are also 'cultures') for a noncriminal putting
to death. Such are the executions of ingestion, incorporation, or introjec-
tion of the corpse" ("'Eating Well,'" 112). Although we will return to the
specifically symbolic register of Derrida's thinking here, what needs to be
emphasized now is that, inasmuch as sacrifice "structures intersubjectivity,"
inasmuch as it is foundational for social and cultural self-definition, it will
be ineluctably tied to discourses of legality, as Derrida argues in "Force of
Law." *The Silence of the Lambs* makes this abundantly clear, for here the law

of culture is explicitly the culture of the law (think how interested the film is in the technologies, training programs, and internal hierarchies of the FBI), and the threat to that culture is explicitly animalized in the totemically named "Buffalo Bill." As a film about the law and its (re)enforcement, then, *The Silence of the Lambs* must arrange its meanings so that there can be a "noncriminal putting to death" of Bill. The first move in such an arrangement is entirely unspoken because unquestioned. It is what seems to be a simple substitution: *We kill animals instead of humans.* But of course, as everyone knows, we do indeed kill humans all the time, and it is in order to mark such killing as either "criminal" or "noncriminal" that the discourse of animality becomes so crucial. We can see how the law of culture arranges its species significations on a kind of grid.

At one end there are *animalized animals.* This pole is, as it were, wholly assumed and is linked to the ongoing practices of violence against non-human others "so vital to our modernity," as Derrida ironically notes ("Force of Law," 951). It is useful here to recall a term broached earlier in this book, the term "speciesism," for it suggests (like its models racism, sexism, and so on) not only a logical or linguistic structure that marginalizes and objectifies the other solely based on species, but also a whole network of material practices that reproduce that logic as a materialized *institution* and rely on it for legitimation. Demme's film, like the "humanism" and "modernity" Derrida critiques, takes for granted the fundamental sacrifice of *nonhuman animals* (in what we eat, what we wear, the testing of the products we buy, etc.), which must continue to be legitimized if the ideological work of marking human others as animals for the purposes of their objectification and sacrifice is to be effective.

Second, there are those *humanized animals*—pets, primarily—that we exempt from the sacrificial regime by endowing them with ostensibly human features. The reach and complicated affiliations of this category will become clear shortly.

Third, there are *animalized humans*, perhaps the most troubling category of all, since all manner of brutalizations carried out by cultural prescription can serve to animalize humans, as can reminders of human beings' mammalian, or even merely bodily, organic existence.

Finally, at the other end, there is the wishful category of the *humanized human*, sovereign and untroubled.

That the ostensibly "pure" categories of "animalized animal" and "humanized human" are the merest ideological fictions is evinced by the furious line drawing at work in the hybrid designations. It is as if these two pure poles can be secured *as* pure (and hence immensely powerful) ideological fictions only by constantly revisiting the locales where they cannot be discerned. That humanism's investment here is more in the ongoing viability of this grid structure than in any specific designation per se is suggested in that the most profound awe will be accorded in the film to the figure who seems to embody *both* poles in their (impossible) purity. In Demme's film, of course, that figure is Hannibal Lecter.

In more general terms, the discourse of animality is present everywhere in Demme's film, chiefly in the central trauma of animal slaughter, which gives the film its title, but also in what feels like the film's organization of the characters by totem clan: Clarice's totem would be the starling ("Fly away, little starling," says Lecter at one point), while her female alter ego, Catherine, is another bird, the martin; Bill is arranged under the buffalo totem by the police in Kansas City, and he also has his own secret personal totem, the death's-head moth; Lecter's totem (fittingly mediated by classical culture) might seem to be the elephant of his Carthaginian namesake, but it is more suggestive to think of it as, perversely, that paradigmatic human of culture, the reader (whose Latin moniker is, after all, *lector*). If anyone is capable of adopting another human as a totem, it is surely "Hannibal the Cannibal."

But while such associations clearly become fanciful at a certain point, the insistence of the discourse of animality and species in the film is not: Dr. Chilton (the bureaucratic psychiatrist who is Lecter's keeper and tormentor) comments to Starling early on that Lecter is a "monster," a "rare example of the species" of pure psychopath, the "only one in captivity." And Lecter will be packaged as such later, when, in a nearly expressionist bit of costume design, he is transported to Memphis in a restraint device that prominently features a mask that has a snoutlike protuberance and tiny bars, like piranha teeth, covering the mouth hole. (It could have been worse, though; earlier Starling, under orders, offers Lecter, in exchange for the name of the killer, a false deal that includes release time on an island used by the government for the study of animal diseases—a place Lecter refers to dryly as "Anthrax Island.") And Lecter's zoological peculiarity is visually reinforced once again when he is held in what looks like a gorilla cage in the middle of a large exhibition room during his time in Memphis, as if on display to the lucky few who have the privilege of visiting this most terrifying and exotic of zoos.

When we turn from Lecter's plot to Bill's, moreover, the category of animalized human becomes even more elaborately figured. The serial killer Buffalo Bill practices a form of animal husbandry on his victim, Catherine Martin; he has a kind of holding pen for her in the dungeonlike basement of his working-class suburban home. Rather than fattening up his animals for the slaughter, he needs to slim them down so he can skin them more easily. He assiduously attends to the upkeep of his animal, and what he does to his victims—selects them for garment-specific characteristics, starves them, skins them, and then stitches their skins together—is precisely what speciesist society does to the nonhuman animal in the practice of ranching and wearing animal fur.

We can see here how Bill's sexual transgressiveness is articulated based on an even more deeply embedded species discourse. FBI wisdom on the motives of serial killers, as Starling dutifully recites it, understands the serial killer as engaged in a kind of "primitive" ritualistic hunting. She surmises that Bill skins his victims as a way of retaining a "trophy" of the kill. But Bill is not so respectful of the law of symbolic substitution. He becomes far more a threat, more "savage," as Lecter puts it, when it becomes clear that the skins are not mere trophies but are being put to use as pure means in the construction of his woman suit. (This is why Bill, when he abducts Catherine, does not sexually assault her but rather is fetishistically focused on her size [fourteen] and on her skin *as skin*.) Bill's goal of having a woman suit is affronting not so much because it expresses an illicit desire to be another sex as because it reveals a confusion about the function of the symbolic nature of "skins" in the law of culture. The problem with Bill is that he does not understand the skins as mere "trophies," that is, as *reminders* of a law of symbolic substitution and hence of the primacy of the symbolic under humanism. The overt artificiality of the trophy, or ornament, its self-advertisement as symbolic, ensures that the continuity between the animal and the human need not be seriously entertained. Unlike the wearer of the fur coat, who "plays" at animality by symbolizing it from a safe distance, Bill thinks his skins will *make* him a woman.

And Bill's confusion about the lines separating animal from human is brought out all the more clearly in that his animalization of Martin is set off against what now can only look like a grotesque humanization of his poodle, Precious. Here again, Bill's sexual transgressiveness coincides with his overdetermination in the discourse of animality. The monstrosity of Bill's twisted relation to a reviled femininity is registered in its full force only when seen as an example of an even more twisted confusion over the line separating human and nonhuman animals. The humanized animal (Precious) and the animalized human (Martin) literally fight for dominance in Bill's basement, and everything depends ideologically on the eventual victory of the animalized human. It is thus of great importance that Martin be seen clutching Precious as she is led from the scene of her horror at the end of the film. The balance has been restored: Precious will now have a proper owner in Martin, who thereby erases her prior animalization by reactivating the potentially perverse pet economy in its proper proportions.

But if this restoration of the difference between animalized humans and humanized animals is one meaning of the scenes with Precious and Martin, the underlying identity of the two is perhaps even more important. For it is clear that in the film's deep-structural working out of the species grid, there are essential homologies between the hostage, the pet, and the sacrificial substitute.[10] In Starling's traumatic effort to save the lambs from slaughter, she tries at first to save the whole lot, and when she realizes they will not take advantage of the opportunity to escape, she picks up one and tries to save it, at least. From among the herd, one is chosen. What we have here, it is important to note, is the principle of symbolic, sacrificial substitution, except in reverse—what could be called the logic of the pet.[11] That Starling's attempt to save the single lamb is not successful, we are led to believe, is what lies behind her commitment to law enforcement and the FBI: the chance to save Martin offers Starling a second shot at saving that one lamb. Structurally speaking, Martin becomes the FBI's pet, as the final image of her clutching Precious advertises.

But the logic of the pet—the sole exception, the individual who is exempted from the slaughter in order to vindicate, with exquisite bad faith, a sacrificial structure she ostensibly protests and belies—extends even to Starling herself. For Starling is clearly the "chosen one" among the herd of FBI recruits. The opening credits sequence has her dutifully submitting to what is explicitly characterized as a regime of (self-) sacrifice ("Hurt. Pain. Agony. Love It!" reads the sign on the training course) when she is plucked from such unpleasantness and installed in the boss's office. Again and again, Starling's "chosen" quality is visually reinforced. A petite Starling steps onto an elevator filled to bursting with red-shirted male hulks, and when she steps out on the higher floor, her "move up" is underscored by the fact that she emerges alone. From among the herd, one is chosen, endorsing the "truth" that People Are Sheep by seeming to oppose it. Hang in there, do your laps on liberal society's cross-country course of hurt and agony, and you too may be chosen.

It is in these terms that we want to specify our differences with a critical discourse that at first seems promising for an analysis of Demme's film. Ecological feminism is likewise concerned with the transcoding of gender and species categories. As noted in the introduction, Carol Adams, in *The Sexual Politics of Meat*, provides a valuable analysis of how the institution of speciesism—and in particular the cultural practice of meat eating—transcodes the edible bodies of animals and the sexualized bodies of women, inscribing both in what Karen Warren calls a shared "logic of domination."[12] (And here we should recall Bill's husbandry of Catherine Martin.) In speciesist, sexist society, both women and animals are subject to a twofold process of objectification through dismemberment (real or figurative) and renaming, a process that foregrounds edible or sexually charged body parts and makes what Adams calls an "absent referent" of the subjectivity and ontogeny of the other. Thus, for example, dead cows are "meat," baby ones "veal," dead and dismembered pigs "pork," and so on. The sexist absenting of women operates by the same sort of renaming of women as animals (chick, beaver, Playboy bunny) and figurative dismemberment (piece of ass, leg man or breast man). These practices, Adams argues, are even more firmly installed in speciesist society because they integrate meat eating with the larger logic of commodity fetishism under modern capitalism, which institutes a radical dissociation between the fetishized act of consumption and the conditions of production absented by it.[13]

While we would surely want to agree with Adams and Warren that the same general structure of "othering" is often at work in the objectification of nonhuman animals and women, the "pet" status and yuppie trajectory of Starling in *The Silence of the Lambs* seem to call for a more nuanced and complex analysis in line with Butler's emphasis on discursive asymmetry. For we must remember that the film calls on Starling to *overcome* her identification with the lambs of her childhood trauma, by means of an energetic ascent out of her working-class past and into the law of culture. It is true that the immediate means for this therapeutic overcoming is the rescue of another woman (Catherine Martin) from brutal objectification. But within the ideological discourse of the film, this compensatory silencing of the lambs only drives a wedge between women and animals, the two homologous objects of Warren's "logic of domination." It says, in so many words, that Starling has finally arrived at full subjectivity because she now understands that lambs cannot be saved, only people can.

In Demme's film, then, the discourses of species, gender, and class are *not* homologous. Indeed, Starling's story remains oppressive in its ideological effect, inasmuch as the opportunity for working-class women bodied forth by her career success is purchased at great expense: a retrenchment of patriarchal law and order, heterosexist gender codes, speciesist relations with nonhuman others, and repudiation of the working class and rural poor in favor of middle-class yuppification. This is why we must see Bill's sexual drama as the negative corollary of Starling's apparently liberatory gender trajectory. In both cases, traditional gender roles are challenged or redefined, but with diametrically opposed consequences. In light of Bill's

transgressions, it becomes clear that Starling's challenge to gender stereotypes is tightly contained within an essentially mainstream ideological
space. Bill's challenge is of a different order. It is not accompanied by class
ascent and yuppification, it is not enabled by the therapeutic reinstatement
of the species barrier, and it does not take place by means of submission to
a bureaucratic, disciplinary regime.

That Starling's seemingly progressive gender trajectory operates in the
service of this larger ideological project is especially clear when we examine
Starling and Buffalo Bill side by side as they enact their struggle in the terrifying, climactic scene in Bill's basement. Bill suddenly reemerges in this
scene as the possessor of that sort of phallic vision that (to borrow Donna
Haraway's phrasing) is "a conquering gaze from nowhere" with "the power
to see and not be seen, to represent while escaping representation."[14] Bill's
startling role reversal, from marked body and seen transsexual (as in his
dance before the video camera) to phallic and unseen male slasher is further
intensified by the night-vision goggles he employs, themselves a form of
those "visualizing technologies" (sonography systems, magnetic resonance
imaging) in which, as Haraway puts it, the feast of phallic vision "becomes
unregulated gluttony" (188–89).

Starling's trajectory in this scene is also suddenly reversed. Blind and
groping, she is faced once again with the unnerving fear that she may not be
"man" enough to get the job done, that she may not ascend to the place of
the father (in this case, *her* father, the dead sheriff) after all but instead may
be violently thrust once more into the place of woman, victim, lamb. Hence
this scene's radical and almost pornographic heterosexism. Bill, who earlier
enacted a spectacular symbolic castration of himself in front of the male
gaze of the video camera in his basement, is now *behind* the lens and is suddenly, terrifyingly male again in the generic role of the secluded rapist-
slasher-killer (as his reaching out to fondle Starling's hair signifies). Starling's blind groping in the frame of Bill's night-vision goggles may thus be
seen as an unexpected analogue to Bill's dance before the eye of the video
camera, one that ratifies Lecter's reading of Bill's sexual drama and discloses
the film's political unconscious of gender. For it is now clear that this—
Starling's blind and helpless groping—is what Bill would have acted like before the camera's eye had he *really* wanted to be a woman.

Starling's impotence in this terrifying scene may be viewed, then, not as
a failure of nerve (for Starling is very nervy, indeed) but rather as a failure—
to put too fine a point on it—of *equipment*. A renounced penis is still a penis,
in other words, and Bill is still, when he needs to be, more than man enough
to handle Starling. At the very last minute, though, the film offers up its

ideological "surprise," as Starling snatches manhood from the jaws of lamb-hood and shows that she does indeed have the necessary phallic equipment (a gun) to become the (and her) father after all. By firmly tethering Starling in this scene to the generic conventions of horror and the "Final Girl," the film thus produces the ideological feint we discussed earlier. Starling—working-class woman who makes good—may still be trying to save the lambs by expanding traditional gender roles, but to do so she is forced to wear the fur, as it were, of the film's heterosexism, a fact thematized by the film's relegating her in this scene to the generic role of horror's "Final Girl."

But the gender discourse at work in this scene must now be reread via the code of species. That Starling's success must clothe itself in the film's speciesism as well is most clear in the final shot at the end of the basement scene. In his earlier scene, before the video camera, Bill's winged kimono re-called the death's-head moth, and in tucking his penis out of sight, he placed himself at the level "below" mammals, as it were, where it is either difficult or ideologically unimportant to draw anatomical distinctions between the sexes. And now that animalizing logic is rigorously pursued to its logical terminus. Bill is represented as distinctly insectlike as he spits and hacks, then lies dead on the floor, large night-vision goggles protruding like bug eyes, hands clenched like mantis claws. A quick cut to an image of the moths and butterflies on the spinning mobile above his dead body hammers home the message: killing this guy has as much significance as squashing a bug. Bill has thus been allowed his mothlike sexual metamorphosis, but it has been mercilessly ironized, all the better to remind us that this is a "*non-criminal* putting to death." And hence the film's heterosexism (Bill as gender bender) is rearticulated at the moment of judgment by its speciesism (Bill as bug). It is as if Bill's failure to recognize that he should affirm the phallus is more profoundly expressed in terms of his failure to recognize that he should be wearing animal, not human, skins. But if anyone deserves the macho nickname "Buffalo Bill," of course, it is finally Starling herself, who comes to reinstate the difference between fur and skin, fag and phallus, white trash and yuppie: the letter of the law.

THE (PSYCHO) ANALYTIC MONSTER

What kind of creature, finally, is Hannibal Lecter? We can start with what we know: "He's a monster," says the odious Dr. Chilton. "Pure psychopath; the only one in captivity." In light of the conventional thinking about mon-sters—that it is their hybrid status, their essence as category mistakes, that makes them objects of alarm[15]—Dr. Chilton's smug assurance that Lecter

represents a kind of "purity" needs to be clarified. For here again we find
something like an ideological feint built into the horror genre. It is undeni-
able, on the one hand, that the monster represents disorder, confusion of
identity, social havoc; everything about the monster's actions and identity
bears this out. But the function of this disorder is to secure, as inverse image
of his own pure otherness, the wishful coherence and purity of the humanist social and ideological order.[16]

But if the monster thus has an unavoidably formal function, he also has a
history, a recent version of which has been elaborated by Slavoj Žižek:

> The pure "subject of the Enlightenment" is a monster which gives
> body to the surplus that escapes the vicious circle of the mirror rela-
> tionship. In this sense, monsters can be defined precisely as the fan-
> tasmatic appearance of the "missing link" between nature and cul-
> ture. . . . Therein consists the ambiguity of the Enlightenment: the
> question of "origins" (origins of language, of culture, of society)
> which emerged in all its stringency with it, is nothing but the reverse
> of a fundamental prohibition to probe too deeply into the obscure ori-
> gins, which betrays a fear that by doing so, one might uncover some-
> thing monstrous.[17]

Žižek here invokes a crucial feature of the homology between the supposi-
tions of psychoanalysis and those of Enlightenment thought (familiar from
Lacan). If one obsession of the Enlightenment project was the interrogation
of origins, traditional psychoanalysis certainly matches that zeal in its own
probing into the original, primal, and primordial. What Žižek calls the "ambiguity" of the Enlightenment—it both opened the passage between the
natural and the cultural, between the human and the "other" and also cease-
lessly elaborated new ways to close off that passage—is also, then, the ambiguity of psychoanalysis. For even as psychoanalysis puts forward an un-
precedentedly complex account of the continuities between the word and
the body, between desire and "fundamental prohibition," between the social
and the "anti"-social, it also inevitably reproduces certain enabling, but
problematic, assumptions about the specificity of the human in relation to
its ostensibly "natural" origins.

Nowhere is this clearer, as noted in the introduction, than in Freud himself, whose intensely imagined primal scenes are as wildly fanciful as anything produced by Rousseau or Herder. Early in his career, Freud theorized
that in the passage from nature to culture, animality to the human, "something organic plays a part in repression."[18] And much later in his career, in
Civilization and Its Discontents (1930), he amplified the thesis considerably,

proposing that early man's learning to walk upright led to "the diminution of the olfactory stimuli" and subsequent disgust at blood, feces, and odorous parts of the body, resulting in a "cultural trend toward cleanliness" and creating "the sexual repression" that leads to "the founding of the family and so to the threshold of human civilization."[19] The essential antinomy of Freud's account, then, is that animality is taken to be both historically continuous with humanity and essentially different from it; and the "missing link" that ensures the disposition of this antinomy—"by a process still unknown," as Freud admits—is "organic repression" (*Civilization*, 52).

Freud's fantasy of origins tells us, then, that the human animal becomes the one who essentially *sees* rather than smells. That dichotomy is powerfully present in the scene of Starling's first visit to Lecter in his cell, when he immediately seizes control of the situation by forcing her to repeat what Miggs said to her—she must repeat the words "I can smell your cunt"—and then sniffing the air being filtered into his cell, only to report that "I myself cannot smell it." What occurs here is both the terrifying resurgence of the animal olfactory (which Freud would claim is "organically repressed") and its effortless surpassing in Lecter's substitution of *culturally* determined olfactory signs—Starling's cold cream, her perfume—for what is coded as her "animal" scent. Lecter here embodies the truth of Freud's fantasy of origin, the same truth the fantasy is meant to disguise—namely, that the Freudian analyst and the Freudian object (here the Freudian animal) are fundamentally coimplicated. For Lecter is the impossible convergence in one body of the analyst and the monster, the one who *sees* (think of the extraordinary power of those tight shots on Lecter's face as he listens to Starling's confessions) and the one who *smells*.

This unsettling and "impossible" convergence in Lecter of analytic vision and animal olfaction makes him precisely what Žižek calls the "fantasmatic appearance of the 'missing link' between nature and culture." Žižek follows Lacan in claiming that the transcendental turn of the Enlightenment (as completed by Kant) consists in the desubstantialization of the subject, its "purification" from its substantial origin in nature, the animal, the bodily, the contingent, in what Kant calls, in *The Critique of Practical Reason*, the "pathological." But if one result of this desubstantialization is precisely the "subject," such a product can never appear without its by-product, what Lacan analyzes under the name of "the Thing" (*das Ding*):

> The subject "is" only insofar as the Thing (the Kantian Thing in itself as well as the Freudian impossible-incestuous object, *das Ding*) is sacrificed, "primordially repressed." . . . This "primordial repres-

sion" introduces a fundamental imbalance in the universe: the symbolically structured universe we live in is organized around a void, an impossibility (the inaccessibility of the Thing in itself). The Lacanian notion of the split subject is to be conceived against this background: the subject can never fully "become himself," he can never fully realize himself, he only ex-sists as the void of a distance from the Thing. (*Enjoy*, 181)

Žižek thus revises the Freudian fantasy of origins by proposing that the Thing never appears as such—as point of origin anterior to the specific codes and structuring force of the symbolic field—but only as an excess or surplus of the symbolic, as the opaque and unintelligible residue that resists symbolization. The subject and the Thing (like the Freudian human and his animal) are, then, coimplicated. The Enlightenment monster, this Thing, expresses exactly the same relation between subject and substance as the Enlightenment subject, *but in the mode of substance rather than in the mode of subject*. Like Ridley Scott's alien, Lecter is less a *thinking thing* than a *Thing* that thinks.[20] And this is why Lecter's horrific mixture of analyst and monster appears most monstrous when it is manifestly allied with the substantial—in sniffing, in eating, most fundamentally, in his *enjoyment*.

What Lecter enjoys most of all, of course, is Starling's trauma. Much of the horror of these "confession" scenes stems from the fact that the structural father substitution at work in Starling's relation to the analyst is crossed by the erotic undercurrent that, only thinly disguised (if at all) informs Lecter's relationship to her. Lecter's desire to know Starling's secrets is clearly a substitute for the sexual knowledge he would like to have of her (as he freely admits at one point), a fact more than underscored by his forcing her to talk dirty, so to speak, by asking her to recount what Miggs said to her. Short of that, Lecter will demand that she make herself vulnerable to him, that she become the object of knowledge for his gaze—in short, that she emotionally disrobe. Demme's direction of these scenes, too, emphasizes their perversion of analytic sessions. While Lecter looks away from Starling (thus adhering to the rule prohibiting eye contact), he does look directly at us, and these tight closeups bring us face-to-face with Lecter's terrifying pleasure in what amounts to Starling's emotional rape.

Finding psychoanalytic paradigms to make sense of these childhood traumas is not especially difficult (as Lecter knows better than anyone). The lambs—and particularly the one she tries to carry away—represent Starling herself, while her desire to protect them symbolizes her attempt to give them the protection she feels she lacks in the absence of her father. On the other hand, as her failure to achieve this protection makes perfectly clear, the lambs also stand for her father himself in his capacity as sacrificial lamb. Lecter well recognizes that in traditional psychoanalytic terms, Starling's desire to protect the innocent by pursuing a career in law enforcement is a blatantly compensatory attempt to resolve the trauma of her father's death that was only reoccasioned by the failure of her original compensatory effort with the lambs. "Do you think," he asks her, "that if you save Catherine Martin, you will stop the screaming of the lambs?" And in the last scene of the film, after the case has been solved and Starling has received her FBI shield, she receives a call from Lecter, who repeats the question: "Have the lambs stopped screaming?"

Starling's silence before this question implies the inadequacy of the compensatory project of her FBI career, and it indexes as well the spuriousness of the psychoanalytic paradigms that define such projects *as* compensatory. The status of psychoanalytic discourse in the film is, as many commentators have recognized, ambiguous at best. Elizabeth Young and Diana Fuss have traced the influence of psychoanalytic paradigms on gender and identity (Young) and on homosexuality and orality (Fuss) to the horror stories the film unfolds; both suggest that the use of psychoanalytic thought in Demme's film is poised between complicity and autocritique. Henry Sussman has read the film as an allegory of the surpassing of a Freudian hermeneutic—"previously our sole comprehensive model of subjectivity" —and its replacement in our postmodern moment by a subjectivity of "emptiness," better analyzed through "the theory of object relations, and the psychopathology of the personality disorders."[21] To the extent that Sussman sees a kind of critical periodizing at work in the film's presentation of psychoanalysis, we would agree, though our perspective on this periodizing, elaborated in the next section, is in a different register.

But Sussman wants, in the end, to psychoanalyze Lecter, and this leads him to conflate Bill and Lecter too readily: both figures, in his view, foreground the "spectacle . . . of the non-mediation of utterly private imagery," what "certain psychoanalysts might call 'primary-process thought'" (167). But if Lecter is an embodiment of primary-process thought, he is so, as we have argued, only in a mode that short-circuits entirely the dichotomy between substance and subject and hence, of course, between primary and secondary process. Lecter's punning and snide invocation of psychoanalytic explanation cannot be reduced to a mere symptom. Primary process inheres in his name: at once reader (Latin *lector*) and licker (German *Lechter*), he can read with surpassing ease all manner of orally fixated lickings; and whatever reading of him you propose, he can lick it. If he embodies primary-process thought, then, he does so knowingly, in the full understanding that such a knowing embodiment *cannot* be explained through psychoanalytic paradigms.

Thus it is that, throughout his encounters with Starling, Lecter implies the paltriness of the models of trauma he so coercively mobilizes against her. In the smile of insinuation that accompanies his unearthing of her "origins" there subsists what Žižek calls the "stain of enjoyment" that clings as an unaccountable residue to every interpretive act (*Enjoy*, 22). Lecter neither believes nor disbelieves the psychoanalytic paradigms he makes Starling bring to visibility; he simply delights in exhibiting their factitiousness. And here trauma and Žižekian "enjoyment" demonstrate their most intimate connection. Once it is recognized that trauma itself is ineluctably fabricated, that it is "the retroactive effect of its failed symbolization" (*Enjoy*, 124), then the remains of that failure, what the originary narrative of trauma cannot wholly subsume, can take on the consistency of enjoyment. From within the symbolic, this failure is experienced as trauma; but from any perspective beyond the symbolic (such as Lecter's), that same failure is experienced as enjoyment.

The basic narrative Lecter forcibly exhibits is familiar from the fourth essay of Freud's *Totem and Taboo*. To kill the father is not to bring about the destruction of the social and its rule of law; it is, on the contrary, to found the social on that very threat and to give birth to that rule precisely as a compensatory gesture. (Lacan will famously extend this mythic narrative in saying that the father is always already dead, that the "name of the father" only retroactively assigns the status of guarantor of the law to the dead father, when in fact the law comes to be only in his sacrifice.) Thus it is particularly fitting that Starling's own submission to the culture of the law also be understood as a reparation, a way to rejoin the phantasmic father of the law (her sheriff dad). When she earns her FBI shield, her surrogate symbolic father, Crawford, speaks the necessary, hopelessly banal words: "Your father would have been proud." Within the symbolic, there are no live fathers, only dutiful substitutions for them, whose mark of fitness is precisely their self-sacrifice, their disavowal of enjoyment, which is distilled, so to speak, in Scott Glenn's wonderfully bloodless performance as Crawford.

But this story of the death of the father, for all that it explains, does not tell all, as Lecter himself knows full well. While it may be the case that "the symbolic order (the big Other) and enjoyment are radically incompatible" (*Enjoy*, 124), it is manifestly not true that there is no enjoyment for the father. Indeed, Lecter's own protectiveness of Starling, his readiness to help her "get ahead," can itself be seen as a kind of paternalistic pleasure, an instance of what Michel Silvestre has called "Father-Enjoyment," which Žižek glosses as the "primordial Thing, i.e., Father-Enjoyment *qua* presymbolic Other" (*Enjoy*, 127). This, then, brings us to the heart of Lecter's threat. In embodying a kind of unavowable "presymbolic other," Lecter exposes symbolicity as such (the assignations of otherness and sameness identified by Derrida) as the core mechanism of Enlightenment and humanist modernity. But in this exposure, it is made clear that Lecter does not respect the principle of the symbolic substitute, the sacrificial victim, the object of exchange, the metaphoric equivalent. Lecter's strategy in the face of these endless substitutions will be to deny their efficacy, to demetaphorize, to literalize, to substantialize. Most momentously, of course, Lecter's cannibalism flouts the originary substitution behind speciesist practice—the killing and eating of animals rather than humans. This deliberate refusal of a symbolic economy is underscored, albeit in an almost offhand way, during a conversation he has with Starling about Bill's "trophy taking": "I didn't keep trophies," he reminds Starling. "No," she concedes, "you ate yours."

It is important to recognize that Lecter's exposure of the hypocrisy of humanist symbolic economies arises not from any kind of resistance to them but rather from his *radicalization* of those very economies, his relentless pursuit of them to their quite logical conclusions. This is clear in his ordering lamb chops (rare) immediately after hearing from Starling about the "screaming of the lambs." Here Lecter uses animal sacrifice not as a symbolic injunction against the killing and eating of humans (as the law of Enlightenment culture would have it) but rather as an *invitation* to it. In ordering lamb, Lecter does not say "I eat animals and not, therefore, humans"; rather, he says "I eat animals *and*, therefore, humans." Moreover, when we recall Lecter's perverse paternalism, this moment emerges as a symbolic violation of the incest taboo, a ravishment or Žižekian "enjoyment" of Starling through eating the flesh of the (Oedipalized and would-be pet) lamb. (That much is made clear, if we missed the point, by the charcoal sketch of Starling, the dress falling from her shoulders, holding a small lamb, that is resting on Lecter's table in the cage in Memphis, just before the chops arrive.)

In our emphasis on the demystifying critical function of Lecter's radicalizing and demetaphorizing of sacrificial symbolic economies, we want to distinguish our account from Georges Bataille's important reading of animality and sacrifice in *Theory of Religion*. Blending Hegel and Freud in a manner somewhat reminiscent of Žižek's reading of Lacan, Bataille argues that the difference between man and animal, individual and thing, subject and object, hinges on the fact that the human, in eating another, engages in a Hegelian "positing of the object as such," through which the subject realizes itself as transcendent, whereas "the animal that another animal eats is not yet given as an object" (and, it turns out, never can be).[22] Bataille recognizes that there is "undoubtedly a measure of falsity in the fact of regarding the animal as a thing" and speculates, in anticipation of Derrida's "'Eating Well,'" that this is why the animal "is fully a thing only in a roasted, grilled, or boiled form"—not only to affirm its object status but also to retroactively "confirm" that "*that* has never been anything but a thing" (39).

"To cut up, cook, and eat a man," Bataille continues, "is on the contrary abominable," even though "it does no harm to anyone" (39). It is precisely this distinction, of course, that Lecter will submit to savage critique, and for two reasons. First, he exposes the factitiousness of the transcendental (and transubstantiating) ideal that drives sacrificial substitution: "Insofar as he is spirit," as Bataille puts it, "it is man's misfortune to have the body of an animal and thus to be like a thing, but it is the glory of the human body to be the substratum of a spirit" (40). Second, in eating his *human* victims raw, not "roasted, grilled, or boiled," Lecter flouts, while knowingly mobilizing, the cultural practices that lamely attempt to paper over the feral, animalistic violence of the *condition humaine* by dressing it up in the attire of gastronomy, religious rite, or, for that matter, Hegelian transcendence (he eats a victim's liver, he tells us, with "a fine Chianti"). "The thing—only the thing," Bataille writes, "is what sacrifice means to destroy in the victim. Sacrifice destroys an object's real ties of subordination; it draws the victim out of the world of utility and restores it to that of unintelligible caprice" (43). This is precisely what Lecter's cannibalism enacts, of course, but without believing in sacrifice's transcendental function or respecting its humanist and speciesist injunctions.

As we have seen, the symbolic law of culture, as it is enacted in the film in Starling's ascent through the ranks of the FBI, concerns fathers only. It is fitting, then, that it is Lecter—that embodiment of "Father-Enjoyment"— who makes the figure of the mother surge forth again in all its disavowed animality. In his startling interview at the airport in Memphis with Senator Martin (the mother of the captive Catherine), Lecter stages the intrication of gender and species (or more specifically, of the feminine and the animal) by recalling this power-suited politician to her essentially mammalian origins in asking whether breast-feeding Catherine had "hardened her nipples." If animals become sacrificial substitutes in the erection of the humanist law of culture, so too does the feminine in the normative ontogenetic narrative of patriarchal culture. "I knew a patient once who kept feeling for his leg after it had been amputated," Lecter digresses threateningly. "Where will you feel it, Senator Martin, when your daughter's lying on a slab?"

Here Lecter evokes an essentially disavowed mammalian union in order to suggest that it is just there, between infant and breast, that a fundamental castration takes place, whose figurative avatars are the man's leg, as well as the animality metonymized as feminine in Senator Martin's breast, and indeed in Catherine Martin herself. After the senator turns away in rage and disgust, Lecter comments, "Nice suit," as if to strip away even further what Kenneth Clark once famously called the "veneer of civilization."

In this invocation of a castrating violence intervening in a pre-Oedipal scene between mother and child, Lecter proves himself once again to be well-informed psychoanalytic theorist. For there is a powerful psychoanalytic account of this castration and subsequent accession to culture, power suits, and senatorial offices. What intervenes between child and mother, what effects the "primordial repression"—of mother, of enjoyment and the Thing, of "nature" and the "animal"—and erects thereby a regime of symbolic substitution and sacrifice, is in fact language itself, or rather symbolicity *tout court*. Once one learns to manipulate a symbolic substitution for the mother—a signifier of any kind, the sound "mom," or a spool, say—one has simultaneously sacrificed the mother and been sacrificed oneself, "hollowed out" (as Lacan would say) by the signifier. Lecter's position with regard to this "primordial repression" effected by the signifier is borne out in his relation to language and the word, which is, as we might expect, one of literalization and demetaphorization. "Look inside yourself," says Lecter, and only clever Starling knows that lurking behind the vapid psychobabble "message" is a material and literal truth: "yourself" is not the sacred desubstantialized interior of the humanist individual, but a storage building outside Baltimore. And most memorable, of course, are Lecter's final words, which offer the best evidence that his literalizing, demetaphorizing strategy is intimately linked to his desire to take the sacrificial economy all the way, to take it precisely at its word: "I'm having an old friend for dinner."

Lecter's relation to language is symptomatic of how "presymbolic others" embody or "thingify" the collapse of the distinction between the symbolic and the Thing, between meaning and enjoyment. This is particularly clear in Bill's practice of inserting moth pupae into the throats of his victims. It is tempting to interpret the moth pupa in rather straightforward Freudian terms as an obviously phallic remainder that expresses Bill's horrible desire for castration (in this sense, it is a reprise of Bill's hiding his penis in the mirror scene). But here we must remember Lecter's acute analysis of Bill's sickness: Bill does *not* desire this castration, finally, but only *thinks* he wants to be a woman. In fact, Bill hates having to choose either affirmation or disavowal of the phallus. Instead, he is living proof of the violence that results from the symbolic gender binary.

Inasmuch as the moth pupa protests and evades a purely symbolic significance, we could see it as an instance of what Žižek calls a "silent scream," a kind of perverse substantialization of the subjective voice, "which bears out the horror-stricken encounter with the real of enjoyment," and which signifies the subject's "unreadiness to exchange enjoyment (i.e., the object which gives body to it) for the Other, for the Law, for the paternal metaphor" (*Enjoy*, 117–18). In aligning himself with the "silent scream," then, Bill testifies to his nonaccession to the symbolic—where reigns, by contrast, the "scream of release, of decision, of choice, the scream by means of which the unbearable tension finds an outlet: we so to speak 'spit out the bone' in the relief of vocalization" (*Enjoy*, 117). In thus placing this scream qua object where it will be found, Bill echoes the perverse substantialization represented by his fashioning the woman suit from the skin of his victims, for both practices indulge in a retrograde reduction of symbol and voice to substantial embodiment, a kind of *"conversion* of the hindered sound into a distortion of matter" (*Enjoy*, 116).[23]

If there is an analogue for Bill's relation to the death's-head pupa, it is surely Lecter's relation to the tongue. For all his articulateness, for all his intimate sensitivity to the tongue as the organ of speech and all that takes place under the aegis of the word, Lecter also knows, in his relentless demetaphorization of all symbolic substitutes, that the tongue is *matter*, that it is good to eat (as he proves when he eats the nurse's tongue directly out of her head without even raising his pulse). Lecter knows that the tongue, for all its detachability, for all its ostensibly "symbolic" equivalence to things such as penises and pupae, is also capable of becoming a "sublime object," an ordinary object "elevated to the dignity of the Thing" (*Enjoy*, 170). In what is one of the more unnerving details of Lecter's monstrousness, we are told that, merely by talking quietly to Miggs for the better part of an afternoon, he somehow managed to get Miggs to swallow his own tongue. In making the symbolic exchanges of an analytic session eventuate in this grotesque substantialization, Lecter offers his most concise lesson in the workings of the analytic monster and gives a rather different spin indeed to the "talking cure."

MIMEMTIC CONFUSION, PERIODIZATION, AND THE PERSISTENCE OF THE AESTHETIC

It has been said that the horror film is the mass-cultural heir of the Greek tragic drama, that aesthetic form most vitally and directly concerned with the foundational ruptures and violences of the social order, with sacrifice and catharsis, and with, as René Girard and Philippe Lacoue-Labarthe have argued, the irreducible intrication of aesthetic representation and mimesis within sacrificial sociality. Lacoue-Labarthe has argued that for the Western self-understanding that has unfolded from the Greeks to the present, mimesis is always a threat, because what is miming and what is mimed are always subject to what he calls the "primal status and undivided rule of mimetic confusion."[24] Who is "like" and who is "same"? Since antiquity, the salient response to the threat of "mimetic confusion" has been sacrificial violence; the scapegoat for Lacoue-Labarthe is always, in some way or other, a *mimos*. But the problem for this "solution" is that this sacrificial violence is itself always a matter of representation, an aesthetic production, itself fundamentally an operation of mimesis. "There is only one remedy against representation," Lacoue-Labarthe concludes, "infinitely precarious, dangerous, and unstable: representation itself. And this is why ritualization and dramatization—the tragicomedy of sacrifice and of the spectacle—never end" (116–17). Lacoue-Labarthe's analysis of the way the aesthetic is in never-ending battle against its own fundamental operation can help us make sense of the bifurcation of monster figures in Demme's film into Bill (who is sacrificed) and Lecter (who is set free). For Bill is clearly the scapegoated *mimos*, representative of "mimetic confusion," the one for whom the symbolic substitution at work in his mimetic fashioning of the woman suit has ceased to be seen *as* mimetic. Lecter, on the other hand, always seems in total control of his own theatrical effects and disguises. Think, for example, of his tourist's outfit at the close of the movie, or of the climactic moment when he removes the ripped-off face of the cop he's been using as a mask. Here the threat of the mimetic confusion seems at once raised to its highest intensity and surpassed, contained, framed by Lecter's very histrionic mastery. If Bill is the sacrificed representative of the social order's response to the dangers of mimetic confusion, then Lecter is the promise (an inevitably hollow one, as Lacoue-Labarthe reminds us) of an aesthetic containment of such threatening mimetism.

That promise is made, in large part, in the familiar terms of what Herbert Marcuse once called "the affirmative character of culture,"[25] which is very much the tenor of Lecter's aestheticism. Despite his radicalization and exposure of all the logics of Enlightenment humanism, what remains largely uncritiqued in and by Lecter is his aestheticism, which is everywhere foregrounded from our very first meeting with him: his charcoal drawings of the Duomo, vaguely reminiscent of Goya; the sketches of Starling we've already mentioned; the copies of *Poetry* and *Bon Appetit* in his Memphis cage; and, of course, the tape deck playing classical music during and after his murder of the Memphis cops. It is useful to recall in this connection that Lecter's sketches and drawings are quite traditional (even clichéd) landscapes and portraits. (No John Cage or Robert Motherwell here!) As the aristocratic bearer of elite culture, Lecter and his aestheticism seem to hold out the possibility of not having to submit, as Starling does and we all do, to the disciplinary regimes of liberal capitalism.[26] Lecter's art and music, like all affirmative culture, compensate him for the freedom he lacks, and when after his escape in Memphis he strings up the cop with a painterly eye toward the Renaissance motif of crucifixion, we are meant to see his aesthetic prowess as the very index of his freedom.

As we know from Marcuse, the freedom promised by the affirmative aesthetic must be read in class terms, as the province of an elite; indeed, it is Lecter's aristocratic class identity that makes his appearance at the end of the film in the bad wig, hat, and sunglasses such a humorous moment. In ideological terms, we could understand this scene as redirecting the freedom we invested in Lecter's class position (via the aesthetic of affirmative culture) into a more fitting class activity for ourselves. Lecter's example of freedom from the middle-class grind turns out, after all, to mean what we always knew it meant: a vacation in the Caribbean, a stroll in a breezy linen suit with hat to match.[27]

But as we have done everywhere in this reading, we need to see how the class discourse here, however ambiguously mobilized, is rearticulated by the more deeply embedded discourse of species. The ideological rerouting we might see taking place at the film's end, in other words, is not accomplished without troubling remainders and leftovers—here the potentially subversive "enjoyment" of Lecter's "animality" that is activated in the bulk of the film and that cannot be sutured closed or recontained by what might be called the *déclassement* of Lecter at film's end. The class discourse so knowingly pushed front and center in the final scene operates as a kind of ideological feint, one that is absolutely crucial to the more profound project of ideological reproduction at work here: the retrenchment of the film's species discourse as the primary means by which the sadomasochistic political unconscious of Enlightenment subjectivity, which links the viewer to Lecter, may be disavowed as "animal" and "perverse." Lecter's *déclassement*, that is, cannot help seeming another instance of his histrionic mastery (we know, after all, that Lecter is still an aristocrat and that his tourist's getup is just a disguise). It thus operates as what Lacoue-Labarthe calls a "theatricalization" that allows us to enjoy Lecter's animality while seeming to contain it in the realm of the aesthetic. The genius of the domestication of Lecter, and hence of the viewer's "enjoyment" generally, is that it cagily allows us to embrace and disavow all at once.

And here, perhaps, is the best place to recall our earlier qualifications about the "ambiguity" of psychoanalytic theory in general: even as it investigates the relation between the natural and the cultural, the ontological and the historical, it finds ever more ingenious ways of closing off that relation by reconstituting historically specific cultural phenomena as productions of an apparently transhistorical social symbolic.[28] The status of Lecter at film's end forces us to revise the standard psychoanalytic paradigm by introducing a finer-grained, periodizing account of the ambiguous movement in which we both repudiate Lecter and allow him to walk off with our utopian dreams. This double movement is, we would argue, the essential—and distinctly *postmodern*—ideological operation of the film. It is not hard, after all, to define the nature of our horror at Lecter; it is more mysterious how we manage so easily to live with him—why, at film's end, we are almost sad to see him go.

Žižek has speculated that one meaning of the distinction between modernism and postmodernism concerns the "status of paternal authority: modernism endeavors to assert the subversive potential of the margins which undermine the Father's authority, of the enjoyments which elude the Father's grasp, whereas postmodernism focuses on the father himself and conceives him as 'alive' in his obscene dimension" (*Enjoy*, 124). In this understanding, the modernist discourse in the film dictates the sacrifice of marginal Bill, while the postmodernism of the film, by contrast, lies in the insistence on Lecter's freedom. For surely Lecter cannot be more succinctly defined than as the "father himself, conceived as 'alive' in his obscene dimension." Unlike modernism, in which we repudiate Freudian animality and primal forces unleashed, postmodernism, in Žižek's account, is characterized by a more ambivalent antagonism toward "the Thing": "We abjure and disown the Thing, yet it exerts an irresistible attraction on us; its proximity exposes us to a mortal danger, yet it is simultaneously a source of power" (*Enjoy*, 122).

We want to emphasize here that Demme's film does not somehow move beyond or dialectically "surpass" the modernist relation to "the Thing"; on the contrary, the viewer's need to repudiate the Thing entirely, à la modernism, is alive and well in the sacrifice of Bill. But by thus siphoning off the viewer's modernist repudiation, the film clears a space to pursue a distinctly postmodern ambivalence toward the Thing, which offers a way of "living with" an unavowable father. Hence the beautiful symmetry of the film's ending: immediately after Starling's boss—a modernist father if there ever was one, wholly drained of his "obscene dimension"—has commended her in the name of *her* father, she is called to the phone and congratulated by Lecter, the postmodern father she can neither banish nor avow. As the figure for the sadomasochistic "Father-Enjoyment" at the heart of postmodernism, Lecter "is the subject's double, who accompanies [her] like a shadow and gives body to a certain surplus, to what is 'in the subject more than the subject [her]self'" (*Enjoy*, 125).

Thus Lecter's cannibalism, which *distinguishes* him from us when viewed from within the modernist-Freudian symbolic—that paradigm that would radically distinguish our meat eating from his cannibalism by invoking the discourse of speciesism—is also the very thing that ties him to us (as what is "in the subject more than the subject himself") when viewed within the purview of postmodernism. The historical specificity of the film is distilled in the fact that Lecter's cannibalism emerges by the film's end as a kind of aesthetic in its own right, a fact more than hinted at in his decadent, opiate lethargy, listening to classical music in the cell in Memphis after eating off the face of one guard and disemboweling the other. Lecter's aestheticism is a version of what Fredric Jameson has analyzed under the rubric of "nominalism," which in the postmodern moment "means a reduction to the body as such, which is less the triumph of ideologies of desire than it is the secret truth of contemporary pornography"—and, we should add, the secret truth of postmodern commodity desire.[29]

The particular usefulness of the periodizing distinction we are drawing here can be explained as follows. The paradigms of Enlightenment (and Freudian) subjectivity will explain the series of bifurcations we have been discussing in the film: of Enlightenment subject into humanist and monster, of monster into Bill and Lecter, and then of Lecter into animal and aesthete. But they will not explain how harbored within this last term—*the aesthetic*—is yet another bifurcation, between the aesthetic mode of affirmative culture, with all the familiar features of class distinction and the promise of freedom and autonomy, and the nominalistic aesthetic mode of cannibalism, which revels in "reduction to the body" and refuses to surrender the object and enjoyment to the other and the law.

Our most obscured, and yet profound, level of identification with Lecter is in this nominalism, and not only in our shared meat eating, which the discourse of speciesism, unmasked by Lecter's cannibalism, would enable us to disavow. For it is this postmodern aesthetic of nominalism and sadomasochistic desire, generalized in commodity fetishism and consumer culture, that is unreadable by the modernist separation of the aesthetic and the animalistic, and that links us to Lecter at a level deeper than the modernist disguises of class discourse and the aestheticism of affirmative culture. By this double movement, the film sets up a strategic misrecognition that shows us ourselves reflected in Lecter precisely where we are not—in class terms, in the tourist on vacation—the better to allow us not to see (and thereby enable us to "live with") the structure of desire that surely binds us to Hannibal the Cannibal. Lecter's cannibalism, then, does not subvert the "official story," the father's story, we might say, of Enlightenment subjectivity, but rather brings it to its logical postmodern terminus in "Father-Enjoyment." Lecter is he who is "in the subject more than the subject himself," the postmodern aesthete who will go all the way with the nominalism and sadomasochism experienced only in diminished form by middle-class consumers of commodities. Unlike us, Lecter won't be having a cheeseburger in paradise. Rather, he will have a friend for dinner.

CHAPTER 4

Aficionados and Friend Killers

Rearticulating Race and Gender via Species in Hemingway

"It's no life being a steer," Robert Cohn said.
—*The Sun Also Rises*, 141

He did not move but his eye was alive and looked at David. He
had very long eyelashes and his eye was the most alive thing
David had ever seen.
—*The Garden of Eden*, 199

HEMINGWAY RECONSIDERED

No modernist writer has come to embody more of the
clichés and caricatures of modernism than Ernest Hemingway, which is another way of saying that no writer is more overdue for a critical facelift.[1] If
"facelift" seems an odd way of sizing up the current situation for modernism's self-styled macho man, it is nevertheless surprisingly appropriate,
for we are beginning to understand that Hemingway—despite the hairy-
chested persona of which he remains the nearly parodic literary exem-
plum—was all along intensely interested in the transgressive possibilities of
gender performativity. In my view, what has made this critical reassessment
of Hemingway possible (and indeed unavoidable) is not so much the ballyhooed recent publication of *True at First Light*—a diffuse and labored piece
of work culled from the vast body of late Hemingway manuscripts—but
rather the posthumous publication by Charles Scribner's in 1986 of Hemingway's unfinished novel *The Garden of Eden*, which he started working on
in mid-1946 and returned to off and on until the end of his life (Burwell, 95,
98). Though unfinished and entangled in a complex textual and editorial
history that scholars are only now beginning to fully understand, this book
is, by nearly all accounts, one of Hemingway's most ambitious and important novels.[2] This assessment hangs in no small part on Hemingway's fascination with themes of androgyny, gender experimentation, and their
relation to creativity, ballasted by what many critics agree is the most

sympathetic and accomplished rendering of a female character (Catherine Bourne) in all of Hemingway's fiction. With *Eden* in hand (and preferably with some knowledge of the larger textual and editorial history), it is impossible *not* to reconsider what has, over the years, hardened into one of the most famous caricatures of literary modernism—a caricature, it should be added, that Hemingway himself did much to establish. West of *Eden*, as it were, what has come into view is a much more interesting and much more ambivalent body of work, one in which, as Mark Spilka has suggested, the Hemingway of Papa's code is not representative of the entire career, or even of its most ambitious undertakings, but is instead anchored largely in the macho posturing—in both life and writing—of Hemingway in the 1930s, a self-commodifying and often desperate chest thumping on Hemingway's part that obscures the quite conspicuous interest in the problematics of gender performativity that bookends that period in Hemingway's work (Spilka, 2).

At first glance, Hemingway's intense interest in cross-gender identification and the transgressive possibilities of gender performativity of the sort he saw in friends such as Gertrude Stein may be seen as his own version of the prototypically modernist rebellion against bourgeois social and sexual mores like that voiced by Ezra Pound in his famous 1917 essay "Provincialism the Enemy." But it does not end there, for Hemingway's rebellion must also be viewed in more pointedly psychoanalytic terms, as a war against the largely unmitigated horrors of living in a universe relentlessly organized by an Oedipal regime of subjectivity—a regime that Hemingway (in prototypically Oedipal fashion) at once loathed and embodied. If we believe Nancy Comley and Robert Scholes in their recent study *Hemingway's Genders*, those horrors are well captured in many of Hemingway's early stories such as "Fathers and Sons" and "Indian Camp," and they are reiterated at the very end of Hemingway's career in *Eden*. Hemingway's texts, as they put it,

> pose the problem of how to attain maturity without paternity. They ask how one can cease to be a boy and become a man without becoming a father like one's own father—and without losing the iterative joys of life. Behind this question, which Hemingway, like Freud, posed almost exclusively in terms of a male subject position, can be discerned dimly the same question posed from the female point of view: How can one become a woman without also, fatally, becoming a mother? If Hemingway's male figures are organized around a problematic opposition of boyhood to fatherhood, his females may well be deployed in a manner that is the shadow of this one, in which the space between girlhood and motherhood is scarcely fit for human habitation. (19)

But if a more probing look at the surprisingly complex and troubled relation between gender and identity in Hemingway's work is making possible
a fundamental reassessment of a major modernist who turns out to be a
good deal more complicated and compelling than his caricatures, the same
cannot be said for another discursive site that is every bit as conspicuous in
Hemingway's work, and every bit as important: the discourse of species.
What this situation has meant—*even* for a writer whose work is so full of
close encounters with animals both wild and domestic—is that the discourse of species, and with it the ethical problematics of our relations to
nonhuman others, continues to be treated largely as if species is always already a counter or cover for some other discourse: usually gender (Spilka,
Comley and Scholes, Burwell), sometimes race (Toni Morrison) or ethnicity (Walter Benn Michaels), still more rarely, class. What I insist on here,
however, is what we might call the *irreducibility* of species discourse and its
problematics—an irreducibility that is especially crucial to understanding
Hemingway. My point is not exactly that we must always decide that species
discourse in Hemingway either is or is not "really about" something else,
but rather that it is precisely that specificity and irreducibility that allows the
discourse of species to do such powerful work in Hemingway's text, to serve
as a crucial "off site"—an/other site—where problems of race or gender
may be either "solved" or reopened by being recoded as problems of species.
As we shall see, that irreducibility is not a "one size fits all" strategy in Hemingway's work but instead functions, I will argue, in diametrically opposed
ways in Hemingway's great early and late novels, *The Sun Also Rises* and *The
Garden of Eden*.

THE STUFFED WORLD: *THE SUN ALSO RISES*

In *The Sun Also Rises*, this crucial irreducibility is made clear from the outset in the foregrounding of two primary elements around which the novel is
organized: the relentless ironizing and, to some extent, androgynizing (as
Spilka has suggested), of the sexual identities of the main characters, Jake
Barnes and Brett Ashley; and the spectacle of the bullfight and the culture
of the aficionado. As for the first, no reader can fail to notice that in this
story of what we might call modernism's ironic perfect couple, the most
masculine and feminine characters in the novel (in terms of traditional, heteronormative gender roles) are at the same time the *least* masculine and feminine—and this, moreover, in opposed ways. Jake, the aficionado, the expert
on boxing and bullfighting, the hard-boiled reporter and skilled trout
fisherman, the seasoned traveler who is at home in bars and restaurants from

France to Spain to New York, is *culturally* the quintessential man's man but is *physically* emasculated because a war wound has left him impotent (with an amputated penis, we are to surmise).

Brett, on the other hand, fulfills the heteronormative feminine gender role in *physical* terms—she is Jake's former lover, of course, and has by novel's end had sexual relations with all the main male characters except Bill Gorton and Montoya. It comes as little surprise, then, when she is figured later in the novel, during the fiesta, as a kind of fertility symbol: "Some dancers formed a circle around Brett and started to dance. They wore big wreaths of white garlic around their necks. They took Bill and me by the arms and put us in the circle. Bill started to dance, too. They were all chanting. Brett wanted to dance but they did not want her to. They wanted her as an image to dance around" (155). And earlier she is cast as a Circe figure who, like the goddess before her, makes men behave like swine (144)—or in this case (as Mike puts it cuttingly to Robert Cohn) "like a poor bloody steer" (142). At the same time, however, Hemingway leans on the fact that Brett is aggressively masculine in both appearance and comportment: she wears a "man's felt hat," is fond of storming into bars and saying things like "Hello, you chaps" (28), and is captivated by the blood and gore of the bull-fight, even as the champion boxer Cohn is squeamish about it (139, 165). And then, of course, there is her famous bobbed hair. Hemingway insists on this ambiguity in our very first meeting with her: "Brett was damned good-looking. She wore a slipover jersey sweater and a tweed skirt, and her hair was brushed back like a boy's. She started all that. She was built with curves like the hull of a racing yacht, and you missed none of it with that wool jersey" (22).

Recent critics have noted how Hemingway's characterization of Jake and Brett complicates the issue of gender identity and its relation to sexual practice and cultural codes. Scholes and Comley find him moving "toward more interesting female characterization" that "preserves the ego strength" of the characters but "justifies their anger or complicates their sexual appetite with other feelings" (42–43). In Brett's case these include feelings such as her refusal to corrupt the young bullfighter Romero to satisfy her own sexual needs (241–43), a refusal that links her, for them, with Jake's stoicism and resignation. And Spilka finds here an example of Hemingway's "quarrel with androgyny," which "again makes us wonder if Jake is not in some sense an aspect of his beloved—not really her chivalric admirer, like Robert Cohn, but rather her masculine girlfriend. . . . What if Brett is the woman Jake would in some sense like to be?" (203–4).

We can return to the passage where Brett is introduced, I think, for a clue to why we should finally be suspicious of this reading. For the question here is not simply the fact of Jake's feminization and Brett's masculinization, but rather the recontainment of that "transgression" within heteronormative gender codes that are far from transgressive. In fact, the more fundamental question is how this discursive site of gender and sexuality gets rerouted through other discursive sites in search of a "solution" to the destabilization of gender that Hemingway has set in motion barely twenty pages into the book. Take, for example, this important passage:

> A crowd of young men, some in jerseys and some in their shirtsleeves, got out. I could see their hands and newly washed, wavy hair in the light from the door. The policeman standing by the door looked at me and smiled. They came in. As they went in, under the light I saw white hands, wavy hair, white faces, grimacing, gesturing, talking. With them was Brett. She looked very lovely and she was very much with them. . . .
>
> I was very angry. Somehow they always made me angry. I know they are supposed to be amusing, and you should be tolerant, but I wanted to swing on one, any one, anything to shatter that superior, simpering composure. (20)

This important early moment may further secure Brett's sexual ambiguity by telling us "she was very much with them," but it also reframes that ambiguity within the context of a larger homosocial bond and its attendant homophobia—the knowing smile of the policeman to Jake, which will be reprised later in the knowing glances and gestures shared by the aficionados—that sends Jake into a fit of violent heterosexual panic. The reason for this panic, of course, is not far to seek, for it is not so much Jake's emasculating war wound as his heteronormative relation to it that overdetermines his reaction.

Focusing, as other recent critics have, on the synecdochic description of the group—the "white hands, wavy hair, white faces"—Comley and Scholes locate Jake's anger in the threat of the potential confusion of his identity with theirs: "He cannot perform, though he desires to do so, while the homosexuals can perform and yet do not desire 'normal' heterosexual sex. The sexually fragmented Jake is thus linked to men he perceives in fragments as unmanly because he has himself been unmanned" (44). Though Jake tries to (re)establish a difference in kind between himself and the gay men by means of the knowing homosocial exchange with the policeman, the novel has *already*, scarcely twenty pages in, given us plenty of reason to be skeptical of such an easy solution. For as Arnold E. Davidson and Cathy N.

Davidson point out, just before we encounter the "boyish Brett and the girl-ish young men," Jake pays for an expensive dinner for himself and the pros-titute Georgette, and on the way to the restaurant, when she puts her hand on Jake's leg and attempts to provide one of her customary services, he pushes her hand away and tells her he is "sick" (15). In the same way that the gay men arrive with Brett and dance with Georgette at the club, Jake is con-cerned here, the Davidsons argue, with "keeping up appearances," even though "the switch in partners suggests, like swinging, the fundamental equivalence of different pairings—Jake and Georgette, Jake and Brett, the young men and Brett, the young men and Georgette. Georgette and Brett (prostitution/promiscuity) are thereby conjoined, and so too are Jake and the boys (sexually maimed/homosexual)."[3]

But the most troubling pairing of all for Jake, I think, is with Bill Gorton. Indeed, Bill is the companion in the novel with whom Jake seems to be hap-piest for longest, and—in an interesting doubling of Jake's relationship with Brett—the novel provides us with a sequence of interludes between Jake and Bill in which Jake's "feminization" is always just under the surface. After Bill arrives in Paris they go out for dinner, and as they walk along the Seine afterward on their way to meet Brett and the others for a drink, we get this scene:

> "It's pretty grand," Bill said. "God, I love to get back."
> We leaned on the wooden rail of the bridge and looked up the river to the lights of the big bridges. Below the water was smooth and black. It made no sound against the piles of the bridge. A man and a girl passed us. They were walking with their arms around each other. (77)

The irony here, of course, is that as the lovers walk by, Jake shares the mo-ment not with Brett, but with Bill—an irony we find repeated later when Jake goes downstairs for breakfast from the hotel room he is sharing with Bill during their fishing trip and hears Bill singing, "Irony and Pity. When you're feeling. . . . Oh, Give them Irony and Give them Pity. . . . The tune was: 'The Bells are Ringing for Me and my Gal'" (114).[4] The point here, of course, is not that Jake and Bill are gay and in the closet, but rather that Jake's wound structurally and, as it were, permanently opens up the possibility of the only sexual outlet that seems to be left to him—that of receptive or "feminine" homosexual partner—in what would otherwise be a prototyp-ical homosocial relationship.

The threat of this sexual connotation always hovers around the edges of Jake's interludes with Bill like the threat of violence that Jake feels toward the gay men in the bar. This fact is by no means lost on Bill. At the fishing hotel, Bill fears he has hurt Jake's feelings by saying, "You don't work. One group claims women support you. Another group claims you're impotent" (115). When Jake tries to play along, Bill says, "Listen. You're a hell of a good guy, and I'm fonder of you than anybody on earth. I couldn't tell you that in New York. It'd mean I was a faggot. That was what the Civil War was all about. Abraham Lincoln was a faggot. He was in love with General Grant. So was Jefferson Davis. Lincoln just freed the slaves on a bet. The Dred Scott Case was framed by the Anti-Saloon League. Sex explains it all" (116). This passage forms a kind of bookend to Jake's moment of heterosexual panic in the bar in Paris, as Bill feels the need to acknowledge that there is something fundamentally unsettling about telling an emasculated male friend how much you like him, which accounts, in turn, for Bill's rapid deflection into outrageous, hyperbolic humor centered on the absurd possibility that paragons of straight white maleness—Grant, Lincoln, Davis—may be something other than they seem. The logic of Bill's joke means that it is just as outrageous to believe that "sex explains it all" as it is to believe that Lincoln and Grant were lovers. Hence Bill's compensatory gesture: not "sex explains it all," but "sex explains *nothing*." This is "common sense" in the same way that the heterosexuality of Lincoln and Grant is—precisely the sort of news, of course, that Jake, given his wound, would like to hear.

As we will see in a moment, it makes perfect sense, within the logic of the book, for Bill to say "I couldn't tell you that in New York" (or, we should surmise, in Paris) because the pastoral setting and its rituals of animal sacrifice are key to securing heteronormative masculine cultural identity and the homosocial bond. For now, however, let me point out that Bill no more believes that "sex explains it all" than that "sex explains nothing." When, in one of the funniest passages in all of Hemingway, Bill makes sport of William Jennings Bryan and the Scopes trial, what becomes clear—it is underscored, as we shall see, in his weird "stuffed animal" excursus in Paris—is that Bill's relationship with animals and what they signify, with how we are linked to them through sex and the body in Enlightenment and Freudian logic, is not nearly so charged as Jake's, because it is not overdetermined by "the wound." For Bill little is at stake, in the end, in negotiating that relationship; but for Jake, it seems, nearly everything—or at least his manhood—is:

"Utilize a little, brother," he handed me the bottle. "Let us not doubt, brother. Let us not pry into the holy mysteries of the hencoop with simian fingers. Let us accept on faith and simply say—I want you to join with me in saying—What shall we say, brother?" He pointed the drumstick at me and went on. "Let me tell you. We will say, and I for one am proud to say—and I want you to say with me, on your knees, brother. Let no man be ashamed to kneel here in the great out-of-doors. Remember the woods were God's first temples. Let us kneel and say: 'Don't eat that, Lady—that's Mencken.'" (122)

Because of the humor here, it is easy to miss the fact that passages like this help secure the fundamental difference between Jake and Bill, a difference that at once is and is not about sex explaining everything or nothing. Bill's joke about William Jennings Bryan (who is referenced both immediately before and after the passage) and the Scopes trial makes fun of those whose religious faith leads them to become unduly exercised about our being linked with nonhuman animals in the evolutionary process. But such a link, as I have already suggested, *is* fundamentally troubling to Jake, for it reminds him of the cycle of sexual procreation and reproduction from which he is barred by his wound.

And on the matter of religious faith, Bill makes a mockery of kneeling in prayer, providing an uproarious lampooning of the Eucharist—*the* central instance of Derrida's "symbolic cannibalisms" that undergird "carnophallo-gocentrism." Here it gets outrageously literalized and brought down to "simian" earth by Bill's injunction against cannibalism—not because the food in question may be of the same species, but rather because eating such a bad writer as Mencken might show poor "taste." All of which may be set in instructive contrast to Jake's sincere attempt earlier in the novel to pray in Pamplona on the way to the fishing in Burguete. There Jake regrets his lapses into a Gortonesque casualness toward the institution of prayer: "I was kneeling with my forehead on the wood in front of me, and was thinking of myself as praying. I was a little ashamed, and regretted that I was such a rotten Catholic, but there was nothing I could do about it, at least for a while, and maybe never, but that anyway it was a grand religion, and I only wished I felt religious and maybe I would the next time" (97). Bill, apparently, wishes he would too, for when he learns after the fishing that Jake was in love with Brett "for a hell of a long time" (123), he says, "Listen Jake, are you really a Catholic?" apparently in the hope that Jake at least has recourse to the *real* Eucharist, and to the transcendence of the body it symbolizes, however faintly (124).

And finally, there is Bill's deflationary posture toward pastoralism and the "temple" of nature, about which it need only be said, perhaps, that while forays into nature are essentially "a good time" for Bill—something on the model of a "vacation" or "camping trip"—they count as a great deal more than that for Jake (and, as we know, for Hemingway). This is clearest, perhaps, in Jake's heavily ritualized handling of the fish he catches, which reads like a passage straight out of early Nick Adams stories such as "Big Two-Hearted River":

> They were all about the same size. I laid them out, side by side, all their heads pointing the same way, and looked at them. They were beautifully colored and firm and hard from the cold water. . . . I took the trout ashore, washed them in the cold, smoothly heavy water above the dam, and then picked some ferns and packed them all in the bag, three trout on a layer of ferns, then another layer of ferns, then three more trout, and then covered them with ferns. (119–20)

The psychoanalytic reading that this passage begs for—the clearly phallic quality of the "firm and hard" trout suggesting that Jake is unconsciously engaging in the ritualistic burial of his own "beautiful" phallus (only, it should be remembered, to be eaten later)—is only reinforced when Bill arrives back with his trout and says to Jake, "How are yours?" and Jake responds, "Smaller":

> "Let's see them."
> "They're packed."
> "How big are they really?"
> "They're about the size of your smallest."
> "You're not holding out on me?"
> "I wish I were." (120)

For those who have spent some time reading Hemingway, it is as if the entire exchange is framed by the infamous "A Matter of Measurements" chapter of *A Moveable Feast*, where Hemingway recounts how Scott Fitzgerald (who was crucial to the composition of *The Sun Also Rises*) complained to him, "Zelda said that the way I was built I could never make any woman happy. . . . She said it was a matter of measurements." To which Hemingway responds—after an impromptu examination—"You're perfectly fine. . . . You look at yourself from above and you look foreshortened. Go over to the Louvre and look at the people in the statues and then go home and look at yourself in the mirror in profile."[5] Of course, it is that very activity—looking at himself in the mirror—that, indirectly at least, leads Jake to the only tears he sheds in the novel as he confronts the graphic reality of his wound (30–31).

That Bill catches the bigger fish is ironically appropriate for an even more significant reason, however. For what matters most, in the end, is not the size or even the killing of the animal, but rather the proper *symbolic investment* in the killing, its *cultural* aspect. From the point of view of Jake's

investment, the point here, we might say, is this: Who needs the penis
when you've got the phallus? Who needs to be exercised about physical
masculinity—or about the relative size of fish—when it's the *cultural* masculinity that counts, a code that secures the transcendence of the humanist *homo* and *vir* (to use Derrida's terms) by *purifying* the subject of that
pathological material aspect that ties it to the world of the Freudian animal and the Kantian Thing? In these terms, the most significant difference between Jake and Bill—of which Jake's attempt to pray and, more important, his investment in bullfighting and *afición* are the most obvious
signs—is that Bill does not believe, as Jake does, in the redemptive power
of the law of culture that is secured, according to Bataille, Derrida, and
Žižek, by the sacrificial symbolic economy that makes possible the transcendence of the human.

This, I think, is the significance of Bill's strange, and strangely insistent,
discourse on stuffed animals when he and Jake are out on the town in Paris.
When they pass a taxidermist's shop, Bill asks, "Want to buy anything? Nice
stuffed dog?" "Mean everything in the world to you after you bought it," he
continues. "Simple exchange of values. You give them money. They give you
a stuffed dog" (72). "Always been a great lover of stuffed animals," he tells
Jake. And later, still more insistently: "See that horse-cab? Going to have
that horse-cab stuffed for you for Christmas. Going to give all my friends
stuffed animals. I'm a nature-writer" (73–74). What is being indexed here is
not only Bill's cynical relation to the commercial values that dominate ur-
ban life in Paris and New York (but not, importantly, the pastoral setting of
the Irati River or Pamplona), about which Jake is similarly sarcastic in a famous passage later in the novel: "Everything is on such a clear financial basis in France. It is the simplest country to live in. No one makes things complicated by becoming your friend for any obscure reason. If you want people
to like you you have only to spend a little money" (233) (or buy a stuffed dog,
if we believe Bill). It is this "Bill-side" of himself (no pun intended) that Jake
fights against throughout the novel, even as it threatens to overtake all of his
relationships, especially with Brett:

> I had been having Brett for a friend. I had not been thinking about her
> side of it. I had been getting something for nothing. That only delayed
> the presentation of the bill. The bill always came. That was one of the
> swell things you could count on. I thought I had paid for everything.
> Not like the woman pays and pays and pays. No idea of retribution or
> punishment. Just exchange of values. You gave up something and got
> something else. (148)

What we learn here is not only the lesson of exchange value, but also that Jake feels sometimes that *he* is "stuffed," insofar as he finds himself in the feminine position of having to "pay and pay and pay," in a kind of endless prostitution that will eventuate in his "pimping" of Brett to Romero, an act that results in his expulsion from the culture of the aficionados near the end of the novel.

Of course, this should come as no surprise at all when we remember the strictly homologous positions of the feminine and the animal in the cultural regime of "carnophallogocentrism." For Jake, to be thrust into the feminine position of having to "pay and pay and pay" is to be emptied of carnophallogocentric subjectivity in the same way that Bill's stuffed animals always already are empty materiality in the culture of *homo* and *vir* that secures its transcendence by their sacrifice. (And here the ironic resonance of Bill's "I'm a nature-writer" cannot be missed in the context of the early Hemingway.) Bill's "stuffed animals," then, are an apt figure for the reduction of the non-human other to brute object, a hollowing out and re-presentation of the nonhuman other that, like taxidermy itself, assumes that the other's existence is always and only for the gaze of the carnophallogocentric subject.

In the terms Sartre uses in his theory of the look in *Being and Nothingness*—a theory contemporaneous with Freud's psychoanalytic discourse on vision in *Civilization and Its Discontents*—the look designates what for me is an other and an object, but that for itself is a subject with its own freedom, for which I, in turn, am an other. To adapt Sartre's theory to the existence of nonhuman others (which, of course, his phenomenology and even more his later Marxism would be unwilling to do), we may say that taxidermy pretends that the look of the other—in this case the nonhuman other—can be reduced to the *eyes* of the other. As Sartre puts it, "If I apprehend the look, I cease to perceive the eyes. . . . The Other's look hides his eyes; he seems to go in front of them."[6] "This," he continues, "is because to perceive is to look at, and to apprehend a look is not to apprehend a look-as-object in the world . . . ; it is to be conscious of being looked at. *The look which the eyes manifest, no matter what kind of eyes they are, is pure reference to myself*" (347; emphasis mine).

The later Lacanian critique of vision will move to triangulate what in Sartre is given as an essentially dyadic confrontation between self and other, freedom and its alienation. What Lacan calls the "scopic drive," like all drives, outruns its instinctual or merely biological underpinnings so that—as Stephen Melville puts it in an especially lucid explanation—"nursing at the breast, evacuating, we engage ontological projects and imagine or reimagine an impossible and unattainable wholeness. These doomed efforts drive us all toward that inevitable situation in which what we take to name the wholeness we demand turns out to be that thing by which society asserts its priority over us and obliges us to acknowledge our incompleteness as unsurpassably our condition. In patriarchy that site is the phallus."[7] The scopic drive thereby covers its own tracks; it makes, as Lacan says, "a tour of its object but its final goal, the site of its imaginable satisfaction, lies in the self" (qtd. in Melville, 15). Lacan's advance beyond Sartre—the "triangulation" I mentioned a moment ago—is to interpose between self and other what Žižek will come to call "the big Other," with a capital O. This is the realm of the symbolic that serves as a kind of material unconscious not locatable in the individual subject (in contrast to the essentially self-transparent subject of Sartre's phenomenology). What Sartre's theory thus misses, as Lacan puts it, is that "in the scopic field, the gaze is outside, I am looked at, that is to say, I am a picture" (qtd. in Melville, 19). "The gaze to which Lacan finds himself exposed," Melville writes, "is not that of another person: it is *outside*. . . . This gaze belongs not to the (small *o*) other but to the Other—language, world, the fact of a movement of signification beyond human meaning" (20).[8]

These terms have a particular resonance, I think, for Hemingway's novel, since Jake's wound so obviously invites a psychoanalytic reading as a figure for how the symbolic order "castrates" or "hollows out" the subject by subjecting him to a profoundly ahuman "outside," which is thematized in the novel by the contingent violence of the war as itself an expression of the castrating "big Other" and the sociosymbolic project. In light of the problem of species discourse, however, what needs equal stress here is that other famous Lacanian term, "the Real," which is underscored by Žižek's rereading of Lacan. As Žižek puts it, the Real is "that which resists symbolization: the traumatic point which is always missed but none the less always returns, although we try—through a set of different strategies—to neutralize it, to integrate it into the symbolic order." The Real is that "which persists as a surplus and returns through all attempts to domesticate it, to gentrify it . . . to dissolve it by means of explication, of putting-into-words its meaning."[9]

I have taken this slight detour into psychoanalysis for two reasons: first, it provides an indispensable background for understanding how bullfighting and the culture of *afición* will function for Jake as something that seems—but finally *only* seems—to separate Jake's ritualized animals from Bill's "stuffed" ones; and second, to sharpen our sense of how a purely psychoanalytic reading of the discourse of species in Hemingway's work would foreclose the possibility of "leaving a space open," as Derrida puts it, for the existence of nonhuman others as anything *other* than a figure for the relation of the symbolic order to the Real or, if you like, the Oedipal subject with its

drives to the body, instinct, and the biological. My point, in other words—
and it is both Derrida's and, less directly, Bataille's—is that the psychoanalytic "outside" of the subject is itself subtended by another, even more re-
mote outside against which psychoanalysis persists in an essentially
humanist effort to secure the human (*even if* that transcendence is rewritten,
as it is in Žižek, as an essentially circular and repetitive failure and, more
important, the castrating knowledge thereof) by relegating the nonhuman
other to the realm of senseless matter, inert organicity, brute instinct, or at
best mindless repetition and mimicry. Žižek is quite candid and quite right,
then, when he insists that "Lacanian theory is perhaps the most radical contemporary version of the Enlightenment" (*Sublime Object*, 7).

Perhaps the best way to state it—to return to Hemingway's novel—is to
say that psychoanalysis in the end interprets the significance of bullfighting
in essentially the same way that Jake Barnes does. Within the Hemingway
code, bullfighting is supposed to symbolize how the uncontrollable violence
of the world (of the war that wounded Jake, of the sexual predation and destructive passion that plays itself out in the fiesta, of the early Hemingway's
own struggle to write sparely and truthfully) can be confronted with
courage, what Hemingway famously called "grace under pressure." All of
these levels and more are in play in Hemingway's famous description of
bullfighting late in the novel: "All that was faked turned bad and gave an un-
pleasant feeling. Romero's bull-fighting gave real emotion, because he kept
the absolute purity of line in his movements and always quietly and calmly
let the horns pass him close each time. . . . Romero had the old thing, the
holding of his purity of line through the maximum of exposure" (168). That
this is a parable of manhood attained through sacrifice of the animal and all
that it signifies in carnophallogocentric culture probably goes without saying, but that is precisely what is so interesting, especially when we remember how the topos of bullfighting has functioned earlier in the novel in the
famous passage where the bulls and steers are unloaded. There Hemingway
uses it to sort the main characters in a way fully consonant with the overarching logic of the species discourse in the novel:

> "It's no life being a steer," Robert Cohn said.
> "Don't you think so?," Mike said. "I would have thought you'd
> loved being a steer, Robert." . . .
> "*Is* Robert Cohn going to follow Brett around like a steer all the
> time?" (141)

The irony here is not that Cohn—the Jew, the outsider—is figured as a castrated male cow over and against the bulls, about whom Jake the aficionado

is most expert and most bullish, but rather that Jake himself, of course, is a steer.

The problem posed by the logic of the novel, then, is how Jake can be a bull without really being one, and how Cohn can be barred from bullishness, even though he is a boxer and like the bull has "a left and a right" and all the requisite equipment. Somehow a difference not just in degree but *in kind* between Jake and Cohn must be enforced, but it ought to be clear by now that this problem cannot be solved on the terrain of sexuality alone. Here is where the irreducibility of species discourse will be crucial to juridically separating Jake from Cohn—and, I should add, to similarly "solving" the problem of Romero's racial identity, which is just the reverse of what we might call "the Cohn problem." For if Jake is linked to Cohn the steer-Jew by his impotence, he is linked to Romero's symbolic bullish masculinity through his *afición*—one whose immediate problem, however, is that it cannot then be seen to privilege Romero's darker race over Jake's whiteness. The "Romero problem," then, is how white Anglo-patriarchal culture can trade on and expropriate the "dark" qualities of the racial other for the purposes of cultural reinvigoration *and* at the same time not elevate that other to a position of ontological superiority—a problem that will take center stage, as we shall see, in *The Garden of Eden*. Romero's race, in other words, must be made to seem beside the point of his *afición*, just as Cohn's sexual performance and physical masculinity must be made beside the point of his cultural identity (or lack thereof). It is this sorting function that the identity made available by the discourse of species (and within that, the mechanism of speciesist sacrifice in bullfighting) will serve.

At this juncture, it might be instructive to contrast my reading with that of Walter Benn Michaels in *Our America*, which views *The Sun Also Rises* as an example of an overarching "nativist" project at work in American modernism, one that is concerned with

> the perfection not of racial identity but of what would come to be called cultural identity. Another way to put this would be to say that the emergence of race as the crucial marker of modern identity was accompanied almost from the start by an acknowledgment of the limitations of race as a bearer of identity—it is these limitations that the technologies of blood supplementation were designed to overcome.[10]

"Blood is blood, but—because blood is blood—blood isn't enough" (12); indeed, if it *were* enough, then identity would be just "a description" and not "an ambition," as Michaels puts it, not a "project" and an "object of desire." "What we want, in other words, may be a function of what we are, but in order for us to want it, we cannot simply be it" (3). Because blood isn't enough in Michaels's reading, Hemingway deploys the "blood supplementation" technology of Jake's impotence to protect the purity of nativist cultural identity from the Jewish Robert Cohn, who, like the figure of the Jew in much high modernism, is associated with fecundity and procreation. As Michaels puts it, "It's as if Jake Barnes were not only 'sterile' in comparison to Robert Cohn but had been sterilized by Cohn" (94). The reason that Hemingway, particularly early in the novel, attends so assiduously to distinguishing between Jake and Cohn is, of course, that Jake and Cohn "are in certain respects so much alike, a fact that the novel makes particularly vivid in their relations to Brett—if, after all, Cohn follows Brett around 'like a poor bloody steer,' it's Jake's footsteps he's treading in" (72).

But what really secures the difference between Jake and Cohn, as Michaels correctly observes, is *afición;* the similarities between Jake and Cohn "are definitively disrupted by the taxonomies of the bullfight" (27) and by the nuances of aficionado culture. Montoya, Jake tells us, "always smiled as though bull-fighting were a very special secret between the two of us; a rather shocking but really very deep secret that we knew about. He always smiled as though there were something lewd about the secret to outsiders, but that it was something that we understood. It would not do to expose it to people who would not understand" (131). For Michaels, *afición* turns out to be another name for "breeding"—you either have it (like Jake) or you don't (like Cohn)—but, paradoxically, you can't have it *just* by virtue of your breeding. ("It amused them very much that I should be an American," Jake tells us; "Somehow it was taken for granted that an American could not have aficion" [132].) *Afición,* then, provides Michaels with an exemplary instance of how the problem of racial discourse is rewritten as a problem of cultural identity, which in turn serves the broader strategy "for insisting upon a race-based model of identity when more literal strategies for preserving it have failed" (13).

Michaels's highly original reading of the novel's problematics of identity seems right to me in many respects, but what is crucially lacking is recognition that this cultural identity is secured and subtended by the sacrificial economy of speciesism. One can't just "be a man" (and for Jake it's a good thing!), because one must *prove* that one is a man; masculine identity is a "project" (to use Michaels's apt characterization), not a given. And the primary way to accomplish that project in Hemingway's novel is to secure the cultural masculinity made available through *afición* by ritualistically purifying itself (through the sacrifice of nonhuman animals) of determination by the purely physical, by "blood" and all that is associated with it (the [maternal] body, the contingent, the sexual, and so on). Blood is blood, and, indeed, blood isn't enough, but not only because it can be polluted by Jews or other racial "impurities." More important, blood always already fatally links us to the world of the animal and the Freudian "organic," to Bataille's "immanence" and Kant's "pathological" Thing, whose surest and most ineluctable sign is our Circelike enslavement to mortality, decay, and the life of the body. This, not national or nativist identity, is clearly where the primary force of Jake's identification lies, and it is why he seeks recourse to the culture of bullfighting—and not, say, to the cultural identity made available by his failed Catholicism—as a way to solve his fundamental dilemma. The issue is not, as Michaels would have it, how to be a nativist "American" without being one (indeed, nothing could seem less important to the novel, and especially to Jake Barnes), but rather how to be a man without being one. The novel is thereby able to turn Cohn's physical and sexual potency into a liability and Jake's impotence into a virtue. The point, then, is not only that cultural identity supplements racial identity in Hemingway's novel, as Michaels argues, but more important, that speciesist sacrifice subtends both. Without the killing of the bulls, there is no *afición* , and without *afición*, there is no cultural identity that can solve the problem of Jake's masculinity *and* at the same time exclude Cohn.[11]

Moreover, paying attention to the specificity of species discourse in the novel makes available a second, crucial dimension of Hemingway's text in a way that Michaels's reading does not: the significance of the homosocial dynamics of aficionado culture. As Comley and Scholes have pointed out, "the framework of the bullring and its culture allowed Hemingway to attend to many aspects of male homosocial desire" and "was appropriated by Hemingway in a manner that allowed him to explore aspects of manliness, including male desire directed toward other males, to an extent that no other cultural context available to him could have provided" (108, 109). Indeed, the knowing glances and physical intimacy among the aficionados are especially hard to ignore: "Montoya put his hand on my shoulder"; "He put his hand on my shoulder again embarrassedly," Jake tells us repeatedly (131). "When they saw that I had afición, and there was no password, no set of questions that could bring it out, rather it was a sort of oral spiritual examination with the questions always a little on the defensive and never apparent, there was this same embarrassed putting the hand on the shoulder, or a 'Buen hombre.' But nearly always there was the actual touching. It seemed as though they wanted to touch you to make it certain" (132).

What is interesting here, of course, is that Jake seems not at all perturbed by this touching of males by other males, even though the emasculation caused by his wound, which led him earlier in the novel to react violently toward the homosexuals in the bar, leads one to expect otherwise. What, exactly, makes this site of male-to-male physical interaction so insulated from unwanted sexual connotations? The Davidsons argue that "the whole ethos of *afición* resembles a sublimation of sexual desire, and the aficionados— serving, guiding, surrounding the matador out of the ring and applauding him in it—seem all, in a sense, steers" (95). Yet it would be more accurate, I think, to say that bullfighting and the culture of *afición* it makes possible is less a sublimation of sexual desire than a *symbolic sacrifice* of it, one that secures the transcendence of the human as bearer of cultural identity. This is why in the bullfight the category of the masculine undergoes a bifurcation into its "cultural" and "natural" aspects (or what Žižek earlier identified as "subject" and "substance," the two faces of Enlightenment identity). After all, we find in the bullfight *two* prototypical symbols of maleness, with the matador Romero embodying its cultural aspect and the bull its physical or "natural" dimension. What is accomplished by this splitting is that the problem of gender identity (or for Michaels, racial identity) may hence be recoded in terms of species identity and the sacrificial law of culture described by Bataille and Derrida: "Insofar as he is spirit, it is man's misfortune to have the body of an animal and thus to be like a thing, but it is the glory of the human body to be the substratum of a spirit."[12] Hence the full resonance of Jake's characterization of Romero's work in the bullring the day after Romero has been severely beaten by the jealous Cohn: "The fight with Cohn had not touched his spirit but his face had been smashed and his body hurt. He was wiping all that out now. Each thing that he did with this bull wiped that out a little cleaner" (219).

This rewriting of the problematic of gender in terms of species is central, for obvious reasons, to Jake's investment in bullfighting; it allows him a way to "be a man" not only in spite of but in some fundamental sense *because of* his sexual failure. But what is equally important, and unavailable in Michaels's reading, is that it also at a stroke cordons off the sexual as such within the sacrificial category of the "animalistic," at once quarantining the significance of Jake's wound for his own masculine identity *and* separating the "homo" from the "sexual" in Jake's relations with the aficionados. Once this is accomplished, Jake's wound may be seen as essentially beside the point of his manhood; and at the same time, the homosocial community of the aficionados is vaccinated, as it were, against the possibility of its mutation into a homo*sexual* community. Once sexuality has been symbolically vanquished and transcended through speciesist sacrifice, men can freely touch, whisper, and engage in other forms of male-to-male intimacy without fear of unwanted sexual connotation. Henceforward one need not fear, we might say, that the "oral examination" will be anything but "spiritual." It is not, as the Davidsons contend, that "Bulls and bullfighters are defined by their sexuality only when they abstain" (96), but rather the reverse. Abstaining is what secures *afición* for Jake and for Romero, and it is only when Romero *doesn't* abstain (in his affair with Brett) that his identity as an exemplar of masculinity is in doubt. And the same is true, of course, for Jake, who in the vicarious sexual activity of pimping Brett to Romero, is ostracized by Montoya from the community of aficionados for introducing the bullfighter to sex.[13]

Mark Spilka thus seems to have it wrong when he argues that in Hemingway's novel "an exchange of sexual roles has indeed occurred" (202) between Jake and Brett, leading us to "wonder if Jake is not in some sense an aspect of his beloved—not really her chivalric admirer, like Robert Cohn, but rather her masculine girlfriend" (203). What such a reading misses, I think, is the crucial maneuver of Hemingway's novel: the way it deploys the discourse of species to secure the heteronormative and homosocial while at the same time ironizing conventional gender codes and exploring possibilities of gender identity that would otherwise be barred. If we pay attention to how the discourse of species operates in the novel, then the reverse of Spilka's reading seems to be true. Brett's maleness stands in the same relation to Jake's as Cohn's does in Michaels's reading. Her "Hello, you chaps," her love of bullfighting, her boy's haircut and her fascination with blood and gore may make her man enough to be a good woman, but she can never be an aficionado precisely because of her "womanly" sexual activity—a fact the novel makes clear not only in her sexual "corruption" of Romero but also in the fact that she leaves the bull's ear Romero gives her in a hotel drawer (wrapped in Jake's handkerchief, no less) with "a number of Muratti cigarette-stubs" (199). She just doesn't understand the fetishistic significance of this token of castration, now transvalued as a sign of the reaffirmation of the culture of carnophallogocentrism (the bull's ear repackaged, so to speak, in Jake's handkerchief).

Brett's distance from the culture of *afición* and the masculine identity it confers, her corruption of it through sexuality, seems borne out by a curious anecdote in *Death in the Afternoon* highlighted in Spilka's reading. There Hemingway writes in an imaginary dialogue,

> The bull is polygamous as an animal, but occasionally an individual is found that is monogamous. Sometimes a bull on the range will come to so care for one of the fifty cows that he is with that he will make no

case of all the others and will only have to do with her and she will refuse to leave his side on the range. When this occurs they take the cow from the herd and if the bull does not then return to polygamy he is sent with the other bulls that are for the ring.

I find that a sad story, sir.

Madame, all stories, if continued far enough, end in death, and he is no true-story teller who would keep that from you. Especially do all stories of monogamy end in death, and your man who is monogamous while he often lives most happily, dies in the most lonely fashion. There is no lonelier man in death, except the suicide, than that man who has lived many years with a good wife and then outlived her. (121–22)[14]

What Hemingway tells us here, in so many words, is that the proper male should not be too dependent on women, still worse on any one woman, because in the end they will all leave you (even if only by having the audacity to die). At best they will, like Brett, leave the bull's ear you give them stuffed in the back of a bedside table with a bunch of cigarette butts; at worst that ear will be *yours*, because you will have been sent to the ring for not learning your lesson about dependence on women and how they, like bulls, must be sacrificed and transcended.

Bataille tells us that "the thing—only the thing—is what sacrifice means to destroy in the victim. Sacrifice destroys an object's real ties of subordination; it draws the victim out of the world of utility and restores it to that of unintelligible caprice" (43). This transformative, almost magical, dissolution of an object's "real ties" is everywhere evident in Hemingway's *Death in the Afternoon*, where he tells us time and again how the animals and events he encounters outside the bullring are utterly transformed once they become part of the pageant. He writes that, even though "I cannot see a horse down in the street without having it make me feel a necessity for helping the horse," "in the bullring I do not feel any horror or disgust whatever at what happens to the horses" (4)—and he has in mind, of course, the terrible goring of the horses that often leaves the animals trailing their own intestines around the ring. This is so, he writes, because "in the tragedy of the bullfight, the horse is the comic character" (6). In "the sculptural art of modern bullfighting," he continues, what happens in the ring is lifted out of the world of the ordinary and the everyday and into a realm of "tragedy," a world of aesthesis where everyday ethical considerations are beside the point. It is the privilege, so to speak, of the animal to serve as sacrificial victim in setting free and memorializing, if only for a moment, the spirit of the human subject to which bullfighting testifies. It is when the bullfighter "moves the cape spread full as the pulling jib of a yacht before the bull's muzzle so slowly that the art of bullfighting, which is only kept from being one of the major arts because it is impermanent, in the arrogant slowness of his veronicas becomes, for the seeming minutes they endure, permanent" (13–14). This is why Hemingway is not bothered by what he characterizes as the "decadence" of the modern bullfight—that the bulls are always new and not allowed to learn from their experience in the ring (21), that they have been bred down in size and are now fought at a younger age than in the past (67–68). Because the bullfight is not a "true sport" (22) or "equal contest" (21) but a "tragic spectacle" (22), Hemingway has no difficulty in asserting that "it is the decadence of the modern bull that has made modern bullfighting possible. It is a decadent art in every way and like most decadent things it reaches its fullest flower at its rottenest point, which is the present" (68).[15]

But bullfighting, of course—and more generally the cross-articulation of speciesism and heteronormative gender in *The Sun Also Rises*—is "rotten" and "decadent" in a more profound sense that we have been discussing in these pages, one that enables us to see that the speciesist sacrifice that secures manhood in Hemingway's novel is not, in the end, a *redemption* of existence from violence through courage, order, and grace—as the Hemingway code would have us believe—but rather a perverse *expression* of that violence driven by a desperate humanism and a panicked heterosexism. From this vantage, violence and the threat of emasculation it harbors are not transmogrified into "tragedy" by the bullfight but are instead *fetishized* by it in a series of desperate therapeutic attempts to remedy, by compensatory domination of a weaker being, the constitutive internal difference of the Enlightenment subject—an internal rift that can never be filled by the very symbolic order that opens it and that promises, sirenlike, to heal it.

As Max Eastman characterized the bullfight in a rather shrill 1934 essay on *Death in the Afternoon*,

> You see these admirable brave men begin to take down this noble creature and reduce him to a state where they can successfully run in and knife him, by a means which would be described in any other situation under the sun as a series of dirty tricks. . . . You see him baffled, bewildered, insane with fright, fury and physical agony, jabbed, stabbed, haunted, hounded, steadily brought dreadfully down from his beauty of power, until he stands horribly torpid, sinking lead-like into his tracks, lacking the mere strength of muscle to lift his vast head, panting, gasping, gurgling, his mouth too little and the tiny black tongue

hanging out too far to give him breath, and faint falsetto cries of anguish, altogether lost-baby-like now and not bull-like, coming out of him, and you see one of these triumphant monkeys strike a theatrical pose, and dash in swiftly and deftly—yes, while there is still danger, still a staggering thrust left in the too heavy horns—and they have invented statistics, moreover, and know exactly how much and how little danger there is—dash in swiftly and deftly and plunge a sword into the very point where they accurately know—for they have also invented anatomy, these wonderful monkeys—that they will end that powerful and noble thing forever.

That is what a bullfight is, and that is all it is. . . . It is not tragic to die in a trap because although beautiful you are stupid; it is not tragic to play mean tricks on a beautiful thing that is stupid, and stab it when its power is gone. It is the exact opposite of tragedy in every high meaning that has ever been given to that word. It is killing made meaner, death more ignoble, bloodshed more merely shocking than it has need to be.[16]

Of course, the author of *Death in the Afternoon* would say that Eastman sounds a bit like the "squeamish" Robert Cohn here, but that is exactly my point. What is much more remarkable, as we shall see, is how he *also* sounds like the narrator of *The Garden of Eden*, the young writer David Bourne, who bears such a striking resemblance to the young author of *The Sun Also Rises* as he is sketched in the Hemingway's contemporaneous *A Moveable Feast*—the narrator who will remember from childhood a three-word pronouncement that will reverberate through Hemingway's *Eden* like a judgment on *The Sun Also Rises*, and on all men who must kill animals to secure their manhood: "Fuck elephant hunting."

NO SYMPATHY FOR THE "DEVIL": *THE GARDEN OF EDEN*

While an intense interest in the problematics of gender identity and its performative dimension is squarely on the table from the very outset of *The Garden of Eden*, what is nearly as prominent, and equally important, is the function of racial discourse in the novel, which intensifies and complicates the investments in the "dark other" and its revivifying primalism that in *The Sun Also Rises* are carried by the figures of Montoya and Romero and, more generally, the culture of Spain, with its robust peasants drinking wine from animal skins on the bus, its all-out festivals of violence and fertility, and so on. But to further complicate matters, the interplay of racial and sexual codes in *Eden* is itself inextricable from a handling of species discourse that is nearly unique in Hemingway and that anchors one of the two main narratives of the novel: not the honeymoon narrative of David and Catherine's relationship and Marita's involvement in it (the very story, we are to surmise, that has become the novel we are now reading), but the "African narrative," the story that David tells of his childhood memory of his father and his father's fellow hunter, the African guide Juma, tracking down and killing an old bull elephant.

At the heart of the present action of the novel is the increasingly adventurous experimentation with sexuality and gender identity introduced by Catherine Bourne in her relationship with David and, eventually, with her lesbian lover, Marita, which eventuates in a ménage à trois from which Catherine herself eventually becomes more and more alienated. As the novel opens, the newlyweds enjoy an idyllic existence on the French Riviera of eating, drinking, making love, napping, and swimming: "There was only happiness and loving each other and then hunger and replenishing and starting over" (14). But it doesn't last long, and only two and a half pages into the novel, Catherine tells David, while sitting over breakfast after making love, "I'm getting hungry already and we haven't finished breakfast" (5). "I'm the destructive type," she continues, "And I'm going to destroy you. They'll put a plaque up on the wall of the building outside the room. I'm going to wake up in the night and do something to you that you've never even heard of or imagined" (5). Scarcely half a page later, we get a hint of what that "something" might be: "They were very tan and their hair was streaked and faded by the sun and the sea. Most people thought they were brother and sister until they said they were married. Some did not believe that they were married and that pleased the girl very much" (6).

Catherine's "something," it turns out, is not only the pleasure she takes in the fact that she and David are mistaken for brother and sister; nor is it only that she soon has her hair "cropped as short as a boy's," "with no compromises" (14–15) so that she and David will be even more indistinguishable, causing a scandal in the small town because "no decent girls had ever had their hair cut short like that in this part of the country or even in Paris" (16). The full force of her "something" is revealed as she sits astraddle David while they make love later that day:

> Then he lay back in the dark and did not think at all and only felt the weight and the strangeness inside, and she said, "Now you can't tell who is who can you?"
> "No."

"You are changing," she said. "Oh you are. You are. Yes you are and you're my girl Catherine. Will you change and be my girl and let me take you?"

"You're Catherine."

"No. I'm Peter. You're my wonderful Catherine. You're my beautiful lovely Catherine. You were so good to change." (17)

To which David responds, thinking to himself as they lay together afterward, "Goodbye Catherine goodbye my lovely girl goodbye and good luck and goodbye" (18).

In time Catherine's explorations will lead her to recruit a lesbian lover, Marita, who quickly becomes sexually involved with David as well. Meanwhile, David's work as a fiction writer centers less and less on the narrative of their honeymoon, as Catherine thinks it should, and more on the African narrative and the story of his father and the elephant hunt. Catherine resents this shift of attention and becomes more and more isolated from David— and from Marita, who is rapidly supplanting Catherine in the role of loving and supportive heterosexual helpmate. Catherine eventually descends into madness, but not before burning the entirety of David's African manuscripts in a crazed effort to force him to return to the honeymoon narrative. David's attentions, meanwhile, center more and more on his isolated negotiation of his relationship with his father through his writing, as he works to complete the African narrative, with Marita—who appears no longer to have an interest in lesbian sexuality—by his side.

At the center of the novel's concerns, then, as many critics have noted, is the relation of gender identity and sexual experimentation to creativity— not only to David's as a writer, but also to Catherine's as an artist of the body and—in her own view, at least—as coproducer of the honeymoon narrative. As Spilka has noted, Hemingway's rendering of Catherine Bourne connects *The Garden of Eden*, like *The Sun Also Rises*, quite directly with the broader interest in androgyny and gender role reversal that we find throughout Hemingway's work, an interest often thematized by what Spilka calls "the hair-matching motif" that we find in (among other works) *A Farewell to Arms, For Whom the Bell Tolls*, and the posthumously published short story "The Last Good Country" (where Nick Adams's sister Litless cuts her hair short to make it look like his) (Spilka, 2, 290–91).[17] "Androgyny" is less the issue here, however, than the dynamics of identification—a point I will return to in some detail later. After all, we find quite different performative dynamics of sexuality and gender in the novel, not all them adequately viewed as androgynous: suggestions of incestuous desire (in Catherine's pleasure that she and David are mistaken for brother and sister); the exchange of gender roles within a still heteronormative code (as when Catherine fantasizes in making love that she has become "Peter" and David is now her Catherine); same-sex desire (as in Catherine's relationship with Marita); and indeed androgyny (in Catherine and David's stereoscopic "merger" toward an identical form of sexually ambiguous appearance). And then, of course, there is David's own identification with Catherine's circulation through these different modes—an identification that seems less about any one of them than about the overarching dynamic they signify: that is, Catherine's unwillingness to be bound by the code of the father and the place it reserves for her.

What makes *Eden* even more interesting, however, is how relentlessly—I am tempted to say how rigorously—it reframes the transgression of both sexual and racial norms in light of the relation between Oedipalism and speciesism familiar to us from Freud (and beyond that, from the biblical narrative of Eden itself). All of which is put right under our noses by the novel's title. For to what, exactly, *does* "the garden of Eden" refer? Does it connote what Comley and Scholes call the "earthly paradise" of "iterative sensual pleasures" from which Catherine's transgression of heterosexual norms precipitates a fall into destructive passion? Here, however, the novel gives us pause in its early characterization of this Eden as "simple"—a word that in Hemingway nearly always means *too* simple. "Don't we have wonderful simple fun?" Catherine asks (10); and two pages later, David muses, "Now when they had made love they would eat and drink and make love again. It was a very simple world and he had never been truly happy in any other" (15).[18]

Hemingway's skepticism about all-too-simple Edens connects him rather directly, and altogether characteristically, to the antiromanticism that is so typical of modernism generally, as in T. S. Eliot ("Sweeney Agonistes," for example), Joseph Conrad (*Heart of Darkness*), and Nathanael West (*Miss Lonelyhearts*), to name only a few. And it suggests one approach to the question, What *is* the status of the "knowledge" that is offered by Catherine in her "devil" mode? by pointing us toward other works of art. To answer that question, we need to remember that the "origin" of Catherine's experimentation with dress, hair, and sexual transformation—an origin that is crucial to the manuscript version but excised from the published novel in 1986—is a famous Rodin sculpture from *The Gates of Hell* titled *The Damned Women*, which represents two female lovers, one of whom appears at first glance to be male, with a short haircut like Catherine Bourne's (Spilka, 286). Even more interesting here, as Comley and Scholes point out, is that one of the titles Rodin used for his massive sculpture *The Gates of Hell* was *The Meta-morphoses of Ovid*, a source text that transmits a warning against homosexual love (in this case, between Iphis and Ianthe): "In all the world of beasts," Ovid writes, "no female ever takes a female" (qtd. in Comley and Scholes, 54). At the same time, however, in Ovid's text, humans can transform their physical identity to match their spiritual state, and it is this moment of transformation that is captured in Rodin's sculpture.

What is most interesting here, then, is not the altogether conventional heterosexist suggestion that heteronormative sexuality is grounded in the very order of nature; nor is it the conventional humanist suggestion that what separates humans from "the beasts" is the human's powerful capacity to transform nature in accordance with desires that are themselves transgressive of "natural" norms and constraints. It is rather the undecidability of the two; the only way to be human is to be *more than* human, to at once embody and transgress humanism's heterosexual code as a determination of nature. At the "origin" of Catherine's desire for sexual experimentation we find, fittingly enough, *no* origin but only an aporia figured as an intertextual instance. The full import of Catherine's "forbidden knowledge," then, is that on the one hand (that of the humanist origin that the Ovidian text warns us to respect) it is "evil," "devilish," and "damned," but on the other (the one that does not respect the constitutive injunctions and repressions that created the desire in the first place) it is the source of creativity, of aesthesis and transformation.[19] In these terms, as I will argue in a moment, the aporetic "problem" of sexual and gender identity that is unsolvable in its own terms (to be human is to at once abide by *and* transcend heteronormative determination) will be rewritten and thereby "solved" in *Eden* as a question not of sexuality but of species, of definitively marking the difference between human and animal.

For now, however, we should note that David's attitude toward Catherine's "devilish" exploits is largely split for most of the novel between exhilaration and remorse. On the one hand, he worries, "What will become of us if things have gone this wildly and this dangerously and this fast?" (21). On the other hand, he muses,

> "You've done that to your hair and had it cut the same as your girl's and how do you feel?" He asked the mirror. "How do you feel? Say it."
> "You like it," he said.
> He looked at the mirror and it was someone else he saw but it was less strange now.
> "All right. You like it," he said. "Now go through with the rest of it

whatever it is and don't ever say anyone tempted you or that anyone bitched you." (84)

At moments like these, David—who has not, we should remember, required very much persuasion to participate in Catherine's experiments in hair styling and lovemaking—seems to be on the verge of registering the productive link between gender experimentation and the kind of negative capability and increased range of identification with the other that is crucial to artistic creativity. As Comley and Scholes observe, "David becomes a better writer as he is led into femininity by his devil. . . . [I]t is certainly the case, whether she is conscious of it or not, that her sexual creativity has stimulated his African writing as well as his writing of the narrative of their life together" (62). (And she is a full partner in another, more literal sense, of course, since it is her family money that has allowed David the leisure to concentrate on nothing but his writing.)

Shortly, however, this creative partnership will become the site of a death struggle between Catherine and David. In a crucial scene that—significantly—takes place before Marita is introduced into the action, Catherine lets slip to Colonel Boyle, an old friend of both David's parents and Catherine's, that she "was a boy in the Prado" where Boyle observed her gazing at a sculpture of Leda and the Swan. After Catherine leaves the Colonel alone with David, Boyle tells him,

> "Remember everything is right until it's wrong. You'll know when it's wrong."
> "You think so?"
> "I'm quite sure. If you don't it doesn't matter. Nothing will matter then." (65)

And then—in one of the stranger pronouncements in the novel—Boyle advises,

> "One small thing more: The get's no good."
> "There isn't any get yet."
> "It's kinder to shoot the get."
> "Kinder?"
> "Better." (65)

The significance of this passage goes well beyond Boyle's advising David not to have children because it will spoil the sensual pleasures he and Catherine have been experiencing on their honeymoon. For we must realize that Colonel Boyle is one of the avatars of the father in the novel. In that light,

the "get" here refers not only to actual children[20] but also any text produced
by the androgynous creative project itself, which Boyle advises David to
abandon, lest it come to threaten his very identity: "Nothing will matter
then." Similarly, Boyle's observation that it's better to shoot the get anticipates the shooting of the elephant by David's father in the African narrative,
which the young David condemns as the act of a "friend killer." In advising
David to shoot the get, Boyle not only aligns himself with the code of
David's father; he also underscores—as we will see in much more detail
later—the structural parallelism between David's cross-gender identification with Catherine and his cross-species identification with the old elephant killed by his father, both of which, Boyle qua father tells us, must be
overcome if David is to enter into full subjectivity.

As the novel unfolds, this struggle will become thematized in the increasing tension between the cooperative venture of the honeymoon narrative and David's isolated work on the African story. David, once energized
and liberated by the sexual experiments Catherine initiates, works more and
more on the African narrative in a separate room he has secured at the hotel. From David's point of view, the African narrative becomes a refuge from
the more and more exciting but unstable world he and Catherine have created for themselves: "You better get to work," he thinks; "You have to make
sense there. You don't make any sense in this other" (146); "remember to do
the work. The work is what you have left" (127). As time wears on, David's
creative endeavor becomes increasingly a locked-room affair with his father,
negotiated in his writing of the African narrative, from which Catherine
must be excluded. While Marita praises David's writerly abilities, and the
African narrative in particular, Catherine deeply resents David's abandon-
ment of the honeymoon narrative and condemns what she sees as his increasingly narcissistic, even masturbatory pursuits (215–16). All of this is
presaged by Catherine's hope that, with the honeymoon narrative, "there'll
never be clippings." After Catherine begins her affair with Marita and introduces her to David, Marita praises David's first book and then kisses him
at Catherine's urging. And when Catherine asks, "Did you think of him as a
writer when you kissed him and liked it so much?" and Marita responds, "I
don't know," Catherine says "I'm glad," "I was afraid it was going to be like
the clippings" (112).

This is only a hint of what is to come, however, for when Catherine finally reads the African narrative, she tears the notebook in two and dashes
it to the floor—"It's horrible," she says, "it's bestial," "I hate you" (to which
Marita responds, "May I have the key, David, to lock it up?" [157–58]).
This in turn anticipates not only Catherine's burning all of the press clip-

pings about David's work (216), but also the catastrophic act of burning all the African manuscripts in an old oil drum behind the hotel (210–11). Catherine, then, is caught in a downward spiral that will eventually lead her to the brink of madness and a final break with David; as she becomes more and more marginalized from David's creative endeavors, she becomes more and more resentful and unstable. Marita, meanwhile, supplants Catherine as the dutiful helpmate, praising and servicing David as an underling rather than challenging him as a partner, as Catherine had done. In response to "the overpopulated vacancy of madness" that has overtaken his life with Catherine, David thinks that "he must go back into his own country, the one that Catherine was jealous of and that Marita loved and respected" (193).

On a deeper level, though, what is going on here is a struggle between Catherine and David's father for David's very soul, a struggle that is mediated—decisively, I will argue—by the significance of the elephant hunting narrative. And here, at the site of this struggle, is where the textual and editorial history of the novel is extraordinarily important, as recent scholarship has shown. As things stand in the version published by Scribner's in 1986, David is in the end reconciled with his father and the code he represents; in the penultimate paragraph we read, "He found he knew much more about his father than when he had first written this story and he knew he could measure his progress by the small things which made his father . . . have more dimensions than he had in the story before. He was fortunate, just now, that his father was not a simple man" (247; see also 146–48). As the published novel ends, David, now wholly served by the dutiful Marita, sits down to recover the African narrative burned by Catherine, and what he discovers is "that the sentences he had made before came to him complete and entire. . . . By two o'clock he had recovered, corrected and improved what it had taken him five days to write originally. He wrote on a while longer now and there was no sign that any of it would ever cease returning to him intact" (247). The message of the Scribner's ending seems clear: David must grow up and move beyond both his foolish gender-bending antics with Catherine and his resentment toward his father's killing of the elephant if he is to become "his own man," not only as a writer but as a person. It is this acceptance of the "complicated" father—and of the sacrifices the code of the father mandates—that enables David to recover the stories, that makes writing not only possible but, indeed, inevitable and unstoppable, as his recovered sentences flow forth in what Catherine would no doubt see as an endless onanistic orgasm: "There was no sign that any of it would ever cease returning to him intact."

But here is where the textual history of the novel is crucial. As Rose Marie Burwell points out in her invaluable reconstruction of the late Hemingway text, the Scribner's novel is culled from over two thousand pages of manuscript now housed in the Kennedy Library at Harvard, and there is no evidence that Hemingway himself ever intended to end the unfinished novel in the manner chosen by Scribner's editor Tom Jencks. As Spilka and Burwell have shown in some detail, however, Hemingway *did* suggest two other endings (Spilka, 208–10). In May 1950, contemplating suicide, Hemingway wrote a provisional ending in which Catherine returns from a Swiss sanatorium, and she and David interact more as patient and nurse than as husband and wife. Marita has dropped out of sight, and David and Catherine agree that if her madness returns, he will join her in committing suicide (Burwell, 105). In 1957–58 Hemingway provided another ending and wrote thirty-nine more holograph pages beyond David's recovery of the African stories. Here Catherine is nowhere in sight; Marita becomes David's "handler" and compares herself to the trainer of a champion race-horse—an ending whose "sad drift," as Spilka puts it, is to "reduce even more seriously 'the girl,' the replaceable androgynous-lesbian muse . . . from the recalcitrant devil who strives for independent creativity at all costs, to the 'good wife'" (310). Added to this is the fact of a very large mirror plot in earlier versions of the novel (deleted by Hemingway) that centers on Nick and Barbara Sheldon, two painters, and Andy Murray, a writer in love with Barbara. Just as Catherine introduces sexual experimentation (and Marita) to her relationship with David, so does Barbara with Nick. Both Barbara and her beloved Catherine are obsessed with looking "just the same" as their husbands, and Barbara falls in love with Catherine, then becomes involved with Andy—a process that ruins her ability to paint but is a boon to Nick's (Burwell, 103).

What the textual history of the novel suggests, then, is that David's affirmative relationship with his father, which seems resolved so unproblematically by the Scriber's ending, is proportionately less a resolution of the struggle between Catherine and David's father's for David's identity than a moment in that struggle—a moment, given David's malleability, that is perhaps not likely to endure. And the weight carried by the Scribner's ending is reduced even more when we remember that the deleted mirror plot shifts the center of gravity of the novel much more toward the complicated relations between creativity and experimentation with gender identity—a shift underlined by Hemingway's May 1950 ending, in which Catherine's contemplation of suicide explicitly refers to that of Barbara Sheldon in the mirror plot (Spilka, 309).

Most crucial of all in reframing the question of David's identity, however, is the elephant hunting narrative. While critics have disagreed about both its meaning and its effectiveness, I will argue here that the childhood trauma of the elephant's murder—in which the cross-species identification common among children is violently foreclosed by the father—may be seen not only as retroactively expressive of, but also as the traumatic *origin* of, David's compensatory cross-gender identification with Catherine and her explorations of otherness in an attempt to break with the sacrificial regime of the father that mandated the killing of the elephant in the first place.[21] For what is immediately striking about the Scribner's version of the novel is how the rather abrupt reconciliation with the father at the novel's end betrays the message registered with sustained and mounting intensity over the many pages that constitute the elephant narrative: that at the origin of David's intense loneliness and sense of isolation from the world and its prevailing codes—a loneliness that leads him to identify with Catherine—is the senseless and brutal killing of an animal.

Just after David begins working on the narrative, we find a passage that condenses many of these associations:

> In the story he had tried to make the elephant come alive again as he and Kibo had seen him in the night when the moon had risen. Maybe I can, David thought, maybe I can. But as he locked up the day's work and went out of the room and shut the door he told himself, No, you can't do it. The elephant was old and if it had not been your father it would have been someone else. There is nothing you can do except try to write it the way that it was. So you must write each day better than you possibly can and use the sorrow that you have now to make you know how the early sorrow came. (166)

The "sorrow now" over the loss of his relationship with Catherine helps him recover "the early sorrow" over the killing of the elephant and the isolation that followed, which in turn led him to identify with Catherine's pursuit of otherness in the first place. I am suggesting, then, that the elephant narrative is not only, as Burwell puts it, "the objective correlative of David Bourne's independent attempt to resist the cultural constraints of family, gender, and race against which Catherine inaugurated their joint resistance in the honeymoon narrative" (99). It also locates animal sacrifice at the very origin of the loss of "Eden" and all that it might signify.

After David and his dog Kibo see the huge elephant in the moonlight

(159), David runs back to the *shamba* to share with his father the wonder of what he has seen. But there he discovers his father and his African guide Juma drunk and making sexual use of the village *bibis* (married prostitutes). Here the father's objectification of the village women parallels the objectification of the elephant that is to come in the hunt. Hence when David's father and Juma discover the trail of the elephant, Juma smiles at him, and he nods back: "They looked as though they had a dirty secret, just as they had looked when he had found them that night at the shamba" (180). "My father doesn't need to kill elephants to live," David thinks:

> I should have kept him secret and had him always and let them stay drunk with their *bibis* at the beer shamba. . . . I'll never tell them anything again. If they kill him Juma will drink his share of the ivory or just buy himself another god damn wife. Why didn't you help the elephant when you could? . . . You never should have told them. Never, never tell them. Try and remember that. Never tell anyone anything ever. Never tell anyone anything again. (181)

As before, David's sense of betrayal of the elephant—and his sense of being betrayed by his father—is at the source of his sense of isolation and loneliness, and it is only made more acute by the fact that as they track the elephant, they discover that he is returning to the site where the animal's *askari*—literally, his friend—was killed by Juma years earlier. Just before the gruesome killing of the elephant is recounted, we read:

> He knew then how much it meant to him to have seen the elephant in the moonlight and for him to have followed him with Kibo and come close to him in the clearing so that he had seen both of the great tusks. But he did not know that nothing would ever be as good as that again. . . . They would kill me and they would kill Kibo too if we had ivory, he had thought and known it was untrue. Probably the elephant is going to find where he was born now and they'll kill him there. That's all they'd need to make it perfect. They'd like to have killed him where they killed his friend. That would be a big joke. That would have pleased them. The god damned friend killers. (197–98)

This passage not only reiterates the main themes of the elephant narrative we have been discussing thus far, it also embeds the symbolic significance of the elephant for David within a larger logic and network of associations that structure the whole of the novel. As several critics have pointed out, the figure of ivory that here links David, Kibo, and the elephant over and against the father also secures the identificatory parallelism of the

elephant and Catherine.[22] "You're just like ivory"; "you're smooth as ivory
too" (169), David thinks (otherwise an odd description, given how darkly
tanned Catherine is), and he muses at other moments on her ivory-colored
hair (156, 178). This association of cross-species and cross-gender identification is anticipated by Colonel Boyle's observation about Catherine: "Do
you always look at them [the works of art at the Prado] as though you were
the young chief of a warrior tribe who had gotten loose from his councillors
and was looking at that marble of Leda and the Swan?" (62)—a characterization that recodes cross-gender escape from the "tribal law" of the male
elders in terms of cross-species miscegenation.[23]

This association is further reinforced in the climactic passage that
records the elephant's death: "The elephant seemed to sway like a felled tree
and came smashing down toward them. But he was not dead. He had been
anchored and now he was down with his shoulder broken. He did not move
but his eye was alive and looked at David. He had very long eyelashes and
his eye was the most alive thing David had ever seen" (199). The narrative
underscores that this is, to David, literally an act of murder by intently focusing on the privileged sensory apparatus of the human in Freud's reading—the eye—as a means of recognizing the being of the nonhuman other.
And the passage moves immediately to deepen and complicate that figure,
first by emphasizing the delicate, long eyelashes of the animal, which link it
via stereotype to the category of the feminine (and beyond that, to Catherine), and second by emphasizing the intensely "alive" quality of the dying
animal's gaze that fixes David—a gaze that is exactly the opposite of the sort
cast by Bill Gorton's "stuffed" animals in *The Sun Also Rises*.

It is important to remember, too, that this cross-species identification is
not limited to just the elephant and what it might symbolize but is extended
to David's dog Kibo as well—a fact that in turn disrupts any attempt to reduce the elephant to merely a symbolic counter for the father or for Catherine in David's Oedipal drama.[24] Nor is David's bond with Kibo simply reducible to "a boy and his dog" all over again, for the power of David's
identification with nonhuman others is that it transgresses the boundaries
between what we have already seen Deleuze and Guattari characterize as "individuated animals, family pets, sentimental, Oedipal animals each with its
own petty history, 'my' cat, 'my' dog," on the one hand, and wild, grand animals like the elephant on the other, "genus, classification, or State animals;
animals as they are treated in the great divine myths."[25] If the figure of ivory
underwrites the parallelism for David between Catherine and the elephant,
between cross-gender and cross-species identification, David's bond with
Kibo across different typologies *within* the domain of the animal emphasizes

the difference or irreducibility of those two forms of identification. In fact, the young David associates his bond with Kibo most of all with the friendship that obtains between the old bull elephant and his *askari* (181).

Unlike David's father, Catherine readily understands this cross-species bond, and her hatred of the elephant narrative is not driven just by her jealousy of it (as David thinks) or by her own frustrated creativity. She criticizes the "cruelty and the bestiality" of the humans in the African narrative and is especially critical of "that horrible one about the massacre in the crater and the heartlessness of your own father" (223). And earlier, before she burns the African stories, she tells David that she never liked his father in the story, "but I like the dog better than anyone except you, David, and I'm so worried about him." "You and Kibo," she continues, "I love you so much. You were so much alike" (163). Then she tells David, after confessing to an especially bad bout of depression in which "suddenly I was old, so old I didn't care anymore": "I'm older than my mother's old clothes and I won't outlive your dog. Not even in a story" (163). By the governing logic of the novel, it comes as no surprise, of course, to find Marita aligned against this set of associations when she asks David if she can read the elephant narrative: "Can't I read it so I can feel like you do and not just happy because you're happy like I was your dog?" (203).

The refusal of the father and all he stands for that is figured by the dead body of the elephant but mitigated by the "girl" Marita in *Eden* is prefigured in Hemingway's weirdly intense story of 1927, "Hills Like White Elephants," by the imaginary dead body of a human fetus whose abortion a reluctant "girl" agrees to consider. Comley and Scholes read this story, and much of the Hemingway text, as posing for the male protagonist "the problem of how to attain maturity without paternity" (18–19). From another vantage, though—the vantage of "the two fathers" and of "Father-Enjoyment" discussed in the previous chapter—this refusal of the place of the father in "Hills" may be viewed as an intensification of its very logic, where the antifather is also in some fundamental sense the *ultra*father. For if you are willing to have your beloved undergo an abortion so that you can continue to travel, drink, and goof off, then you *already* resemble the father of *Eden*, with his hunting trips, drunken sprees, and bare tolerance of the presence of his child around the fringes of both—the father whose law is "perverse," to use Žižek's term, not only because that law is instantiated ex nihilo, across a void, but also because it mandates the sacrifice of the very Thing (as fetus) that is also the object of enjoyment (as the female body). As Žižek puts it in his reading of the fetal "desperate homunculus head" of Edvard Munch's *The Scream*,[26] the perversity here is not so much the "worn-out commonplace" that the subject is "barred and barren, crossed out" (138), but rather that "the limit impeding the subject's self-expression is actually *the subject himself*" (137), who exists as "*nothing but* this dreaded 'void'—in *horror vacui*" of distance and failed disavowal "from what is 'in him more than himself,' from the Thing in himself" (137–38). What "Hills" enacts, then, is a kind of unnerving literalization of the sacrificial economy of the Enlightenment subject thematized in *Eden,* but one whose radicality is that it *does not* respect the symbolic substitution of animal for human as the victim of Derrida's "noncriminal putting to death." It is as if the young man in "Hills" is David's father in *Eden* enacting the fatherly advice Colonel Boyle's will give to David—"The get's no good," "it's better to shoot the get"—thus providing a kind of prolepsis of David's speculation that his father would kill *him* and Kibo too if they had ivory.

"Hills," then, demystifies the speciesist sacrificial economy of humanism precisely by *not* respecting the juridical function of species difference in sacrificial violence. Moreover, in the "girl" "Hills" provides a figure who demystifies even as she suffers that symbolic economy's radicalization. For when the boy imagines the freedom such sacrifice will bring and says, "We can have the whole world," the girl replies, "No, we can't" (276). And when he continues to push, she responds with the threat "I'll scream," one that is beyond words and that links her via the sheer materiality of sublinguistic vocalization to the disavowed "animal" world with which she is cognate in humanism's symbolic logic. To borrow once again from Žižek's reading of Munch's painting, what we find here, however, is a "silent" or "hindered" scream that "cannot burst out, unchain itself and thus enter the dimension of subjectivity" (117) and thus "finds an outlet (one is tempted to say 'acted out') in the anamorphotic distortion of the body" (116)—in the case of "Hills," in the distortion of her body by the fetus-Thing within.

"Hills Like White Elephants"—with its lack of narrative direction (embodied in the setting of the train station with the tracks stretching out barrenly in both directions) and its exposure of Oedipal authority and self-congratulatory humanism—may be read as an unsettling and beguiling *formal* deployment as well that is instructive for our understanding of *The Garden of Eden.* As Judith Roof argues in *Come as You Are: Sexuality and Narrative,* in Freud sexual perversion and narrative perversion are twinned concepts.[27] "Supplanting the proper conclusion"—that is, heterosexual "discharge of the sexual substances," as Freud puts it—"perversions cut the story short, in a sense preventing a story at all by tarrying in its preparations. But this premature abridgment only has significance in relation to the 'normal'; we only know the story is cut short because we know what length the story is supposed to be" (xxi). In these terms, what is "perverse" about "Hills" is precisely that it has taken heterosexual humanism at its word—all too faithfully, as it turns out. The "proper discharge of the sexual substances" has eventuated in the prospect of its own reproduction via offspring, which must now be cut short in a radical exposure of a symbolic economy whose mere "alibi" is the family narrative and caring for others: all along it really *was* about the absolute freedom of the father—to goof off, to kill and eat what he pleases, to be wholly unencumbered by women and children. In this light, David in *Eden*, in the process of writing and rewriting the African narrative, moves ever closer to fulfilling what Roof characterizes as "narrative's trajectory toward mastery" (148)—of himself, of the unruly and "wrongheaded" participation in Catherine's experiments, of his "childish" cross-species identification with the elephant. What Roof calls "narrative's transformative dynamic that constantly converts disorder into order and mastery" is surely on display here, for in the process of recovering and rewriting the "improved" narrative of the elephant hunt, David forgives his father and regains Oedipal mastery over his own identity, with the geisha-like Marita at his side.[28]

If we trace backward the narrative process David follows, however, we find that David's rather abrupt reconciliation with his father and his reinstatement of Oedipal, heteronormative narrative belie the radical "disorder" (to borrow Roof's terms again) voiced in the African narrative that *we* read. Indeed, we find instead a violent trauma of betrayal and murder so painful and intense that David cannot even bring himself to sit down and put it into words for quite some time—not, that is, *until* he is enabled by participation in Catherine's "perverse" creative project. When David puts aside the honeymoon narrative to begin work on the elephant hunting story, he thinks that the latter "had come to him four or five days before" (93); then he realizes "how long he had intended to write it," that "the story had not come to him in the past few days. . . . It was the necessity to write it that had come to him" (93). Fifteen pages later, he writes "the first paragraph of the new story that he had always put off writing since he had known what a story was" (108), and then, after another fifteen pages, it is characterized as "attacking each thing that for years he had put off facing" (123).

IDENTIFICATION AND ITS DISCONTENTS

It makes perfect sense, of course, that Catherine is in some profound way responsible for David's ability to finally confront the father and his code, for she enables David's cross-gender and cross-species identification in a process, to borrow Diana Fuss's phrasing, that would "place the traditional subject of knowledge in the unsettling position of object"—as David the child does by condemning his father as a murderer, his sexual use of the *bibis*, and so on—"calling into question the very boundary that would render the human transparent while subjecting its imaginary others to relentless scrutiny and obsessive reclassification." As Fuss reminds us, "The vigilance with which the demarcations between humans and animals, humans and things, and humans and children are watched over and safeguarded tells us much about the assailability of what they seek to preserve: an abstract notion of the human as unified, autonomous, and unmodified subject."[29] The attempted recontainment of this demystification by an Oedipal mastery that turns out to have been always already accomplished (if we believe the narrator) is interesting, however, not so much for its success as for its *failure*. There is no better example of this dialectic of demystification and recuperation in *Eden*, perhaps, than the narrator's observation about David's cross-species identification in the elephant hunting narrative, that "it was a very young boy's story, he knew, when he had finished it." This desperate attempt to recontain the child's identification with the animal and his suffering has been shared, I should note, by more than a few of Hemingway's critics. E. L. Doctorow, in a famous review, says of the elephant narrative that it is "bad Hemingway, a threadbare working of the theme of a boy's initiation rites" (qtd. in Spilka, 301). And even Spilka—who wonders about Hemingway's "long-buried confession of deeply troubled ambivalence about such 'friend-killing'" (189)—treats the elephant narrative as essentially a resurgence of Hemingway's "boyhood romanticism" (189). In both cases, of course, what we find is an inability to take seriously the subject position of *either* David the child *or* the animal whose killing is so traumatic, an inability that rigorously obeys the Freudian logic telling us that such "zoophilia" is the mark of the not-yet-subject and is indeed "perverse" if it persists into adulthood.[30]

As I have already noted, what makes this "unsettling" confrontation with the traumatic elephant hunting experience and with his father possible is the space opened up by David's "perverse" creative partnership with Catherine. *At the same time*, however, this cross-gender identification may be seen as itself rooted in and generated by the even more "perverse" cross-species identification he experiences as a child. Because the relation between cross-species and cross-gender identification is both structurally parallel and, in the novel, temporally circular—the latter is rooted in the former, but the former becomes accessible only because of the latter—any idea of an "origin" or "repressed truth" in David's experience must be abandoned. Far from compromising the critique of the humanist symbolic that the novel makes available, however, this fundamental retroactivity or circularity helps to secure it, for if this *absence* of origin—which the humanist symbolic rewrites as a *lack* of origin—can be shown to be coterminous with the problematics of subjectivity as such, then it is the humanist attempt to imagine its own origin, to be present at its own (self-) birthing, at and as a moment of rupture with the animal and the organic (as Freud does in *Civilization and Its Discontents*) that is exposed as a fantasy and a childish, sentimental bit of anthropomorphism, not the "childish" cross-identifications of David.

Of course, the fundamental circularity and retroactivity that I have been discussing might itself be seen, if we believe Lacan and Žižek, as the ironic deep structure of the Enlightenment schema. From a Lacanian-Žižekian point of view, the "truth" of David's aloneness, in other words, would reside not in this specific instance of the killing of an animal, but rather in the more generalized economy of subjectivity and the workings of the symbolic as such, which are themselves founded on a sacrificial symbolic economy thematized and re-presented here in David's "experience" *as* a traumatic "origin." From a Lacanian-Žižekian point of view, then, what might be on display in *Eden* is the difference between "symptom," conceived in Freudian terms, and what Žižek, following Lacan, calls the "sinthome." In classic psychoanalysis, "the symptom is always addressed to the analyst"—whose role in the novel is filled, of course, by the father-narrator: "It is an appeal to him to deliver its hidden meaning. We can also say that there is no symptom without transference, without the position of some subject presumed to know its meaning" (*Sublime Object*, 73). The sinthome, on the other hand, testifies to the persistence of the traumatic Real or Thing in the symptom that "resists symbolization absolutely" and that cannot be dissolved through interpretive or analytic work. Indeed, it provides the alibi for such work and for the various codes of the symbolic—such as the father-narrator's Oedipalism—that attempt to "gentrify" it (69). As Žižek puts it, the sinthome is thus "an inert stain resisting communication and interpretation, a stain which cannot be included in the circuit of discourse, of social bond network, but is at the same time a positive condition of it" (75).

In this light David's experience might be seen, then, as exemplary (to use Rey Chow's phrase) of "the fundamental misrecognition inherent to processes of identification" (qtd. in Žižek, 66), insofar as what is disclosed in David's identificatory encounter with Catherine and with the elephant is not "the truth of the other—what he or she really is" (70), but rather the way the fantasy of such a truth covers over the failure of the symbolic to confer consistency upon the subject, a failure to dissolve and render meaningful and transparent the traumatic Real or Thing that is "in the subject more than the subject himself." The essentially Freudian and Oedipal recontainment of these identifications by the code of the father-narrator would thus be revealed as a failed symbolic attempt to cover over a deeper, more traumatic psychoanalytic truth: that "the name of the Father" runs aground on what Žižek calls the "rock" of the Real (69) as bodied forth in the traumatic encounter with the feminine and animal body now coded as the Thing, *das Ding*.

Precisely here, however, is where we may locate the limits of a psychoanalytic reading, and specifically of the concept of identification, for understanding *The Garden of Eden*, *even if* those are rewritten in Lacanian-Žižekian terms. I am not suggesting, of course, that we should or can do without the psychoanalytic reading and its paradigms—indeed it seems impossible to me to make much sense of *The Garden of Eden* without them. But I *am* saying that the ethical as well as theoretical significance of the elephant hunting narrative and of species discourse in this novel cannot be accounted for within the terms of psychoanalysis alone. We have already seen in the previous chapter how these limits manifest themselves in Žižek's own work; but to put it another way here, what joins and illuminates the two primary instances of cross-identification in *The Garden of Eden* is not some sort of *ontological positivity* that binds David to the feminine and the animal via the category of Real, as Žižek would argue (*Sublime Object*, 75).[31] Rather, it is a *formal, structural effect of negativity*, produced by the rigorously systematic force of exclusion that makes the *place* of David the *child*, of the elephant as *animal*, and of Catherine as *woman* "the same" and yet "different": the same in their formal isomorphism as "outsides" vis-à-vis the humanist symbolic, but different insofar as the discourses of "child," "animal," and "woman" bear differentially distributed effects of that formal exclusion in the social and material terms. It is precisely in this "sameness in difference," and in a materiality that is not, I am tempted to say, "merely ontological," that the ethical force and complexity of the novel resides. The "difference" of these different instances of identification can be preserved, in other words, precisely to the extent that they are *not* collapsed into epiphenomena of an ontological positivity via the Lacanian Real—precisely to the extent, to put it another way, that we are willing to insist on the differences between ontology, discourse, and institution.

It is a question, as Judith Butler has recently pointed out, of the ability of "the lost and improper referent" to speak, and it is a problem, of course, not only for species difference but for gender and sexual difference as well. As Butler puts it in her critique of Žižek,

Paradoxically, the assertion of the real as the constitutive outside to symbolization is meant to support anti-essentialism, for if all symbolization is predicated on a lack, then there can be no complete or self-identical articulation of a given social identity. And yet, if women are positioned as that which cannot exist, as that which is barred from existence by the law of the father, then there is a conflation of women with that foreclosed existence, that lost referent, that is surely as pernicious as any form of ontological essentialism.[32]

So if we want to continue to use the term "identification," then, we must realize, as Fuss has argued in *Identification Papers*, that identification "operates as a mark of self difference, opening up a space for the self to relate to itself as a self, a self that is perpetually other" in "a process that keeps identity at a distance, that prevents identity from ever approximating the status of an ontological given."[33] But if identification is "from the beginning, a question of *relation*, of self to other, subject to object, inside to outside" (3), then the problem with the concept of identification in *Eden*—a problem that finally reveals the most radical and challenging discourse in the novel to be not of gender but of species—is that this "relation" in David's cross-*gender* identification with Catherine remains framed within an essentially *representational* symbolic economy. If the question for the concept of identification, as Fuss candidly admits, is "How can the other be brought into the domain of the knowable without annihilating the other *as other*—as precisely that which cannot not be known?" (4), then what makes cross-gender identification finally a kind of feint or "lure" (to use the Lacanian term)[34] in Hemingway's novel is the essentially *mimetic* form it takes, readily thematized in the novel by the "androgynous" matching of hair length and color, clothing, tans, and so on that so many critics have noted. For the problem with the mimetic recontainment of the identificatory dynamic, of course, is precisely its identitiarian reinstatement of a relation of adequation and transparency between self and other, original and copy, and so on, and hence its effacement of the alterity of the signifying instance that Fuss is rightly concerned to preserve.[35] To put it another way, it is the overwhelmingly mimetic mode of the cross-identification between David and Catherine that threatens to rewrite what Fuss calls "mimicry" (the "deliberate and playful performance of a role") as "masquerade" (a "nonironic imitation of a role") by collapsing the distinction between the two—which "depends on the degree and readability of its excess"—into a representationalist fantasy of the sign that says what one looks like is immediately expressive of what one *is* (*Identification Papers*, 146). This mimeticism is politically and ethically problematic, of course, because, as Fuss puts it, "the signifier 'Other,' in its applications, if not always its theorizations, tends to disguise how there may be Others—subjects who do not quite fit into the rigid boundary definitions of (dis)similitude, or who indeed may be left out of the Self/Other binary altogether" (144).

That space is signified in Hemingway's novel, as I have been suggesting, by that other object of cross-identification so conspicuous in the novel: the *nonhuman* other. It should come as no surprise, then, that David's identification with the elephant is presented in anything but mimetic terms, as he cycles through a set of irreconcilable descriptors of the animal that seem to trace an identification at once intimate and exterior. The elephant is described, variously, as smelling "strong but old and sour" (159), as a "friend" (180–81), a "brother" (197), "gray and huge" (199), "the most alive thing David had ever seen," having "very long eyelashes" (199), his "hero now as his father had been for a long time" (201), and after he is killed, as a "huge wrinkled pile" (201). Whatever the elephant signifies in David's cross-identification, it seems little help to think of it in terms of a mimetic relation, and indeed its signifying force seems to reside more in its movement along a signifying chain that traverses the borders of David's world without condensing into any single association—borders to which the young David himself feels more and more consigned. If, as Fuss suggests, "Psychoanalysis erects itself, as a science, *against* the literary" by repudiating a "power of figuration" it cannot do without—and if, moreover, "Metaphor, *the substitution of the one for the other*, is internal to the work of identification" (*Identification Papers*, 5)—then we might say that the greater power of cross-species versus cross-gender identification in *Eden* is that it prevents the *condensation* of this metaphoric principle into an ossified mimetic one, and in so doing it resituates each metaphoric instance of identification within the larger movement of a metonymic signifying chain.

Perhaps the best way, then, to view how the novel handles these issues is to say that David's relationship with Catherine invites the psychoanalytic characterization of "identification," but the "difference of the difference" here is that that cross-gender "identification" is a symptom (or sinthome, in Žižek's terms) of what is *radically other than identification itself* in the young David's relation to the elephant and that therefore functions as the more adequate—because less representational—figure for the essentially nonrepresentational dynamics of cross-identification as such. In this light, David's identification with Catherine may be seen as rescripting and in some sense domesticating, via *psychoanalytic* paradigms, the more radical "identification" with the elephant in the same way that David's final resolution with his father only further domesticates the psychoanalytic dynamics of his relationship with Catherine by rescripting *them* as always already Oedipal. We thus find in the novel a set of expanding, concentric "outsides"—from David's narcissistic identification with his father via the "locked-room" work on the African narrative, to the mimetic cross-gender identification with Catherine and her rejection of traditional gender types and sexual practice, to the homologously external place of the child, and finally to the animal and to cross-species "identification," whose force must finally be read outside the limits of psychoanalytic paradigms altogether, even as it cannot wholly escape them.

To say as much, however, is to remain still, perhaps, within an essentially humanist discourse. For as Derrida has recently pointed out,

> Beyond the edge of the *so-called* human, beyond but by no means on a single opposing side, rather than "the Animal" or "Animal Life," there is already a heterogeneous multiplicity of the living, or more precisely (since to say "the living" is already to say too much and not enough) a multiplicity of organizations of relations between living and dead. . . . These relations are at once close and abyssal and they can never be totally objectified. They do not leave room for any simple exteriority of one term with respect to another.[36]

Indeed, here is where the force of cross-species identification in the novel might well be read under the *anti*psychoanalytic sign of Deleuze and Guattari. In the fascinating sections on animality in *A Thousand Plateaus* that we have already touched on, what Deleuze and Guattari call "becoming-animal" "always involves a pack, a band, a population, a peopling, in short, a multiplicity" (239). What Deleuze and Guattari aim to underscore is that the animal properly understood is a privileged figure for the problem of difference and subjectivity generally, because it foregrounds how the subject is always already multiple. Or as Donna Haraway puts it in a complementary passage, "One cannot 'be' either a cell or molecule—or a woman, colonized person, laborer, and so on. . . . We are not immediately present to ourselves."[37] Thus "the topography of subjectivity is multi-dimensional. . . . The knowing self is partial in all its guises, never finished, whole, simply there and original" (193). From this vantage, the particular force of cross-species identification in Hemingway's novel might thereby be located in the child's enactment of a "becoming-animal" that radically destabilizes not only the "molar" "diagrams" of subjectivity (to use Deleuze and Guattari's terms) that constitute the code of the father, but also the domestication of the radical multiplicity of gender and sexuality into mere "swinging," we might say, via the mimetic principle that finally arrests the novel's cross-gender identificatory dynamics.

As Brian Massumi puts it, for Deleuze and Guattari "the proliferating metaphysical splits between otherness and identity, Imaginary and Symbolic, signified and signifier, subject of the enunciation and subject of the statement, translate a real bodily bifurcation: between the human person and its subhuman individuals," including the "subhuman" evolutionary history that the human subject shares with animals like elephants and the entire order Mammalia. This "multitude of individuals that contract to produce the person [under Oedipalism] is reduced to the one-two-(three) of self-other-(phallus), distinctions which can exist only on the second-order level of identity and identity loss"—or, for our purposes, on the level of the mimetic identificatory relationship between David and Catherine.[38] Indeed, from a Deleuzean-Guattarian vantage, the surest sign of the novel's humanism would be not so much David's "anthropomorphic" cross-species identification with the elephant as the pet economy that, for Deleuze and Guattari, would Oedipally recontain David's relationship with his dog Kibo, and would, beyond that, determine David's inability to establish any ethical linkage between the multiplicity of animal others he encounters in the novel, such as the "two spur fowl" David kills with his slingshot while they track the elephant (172). The Derridean "heterogeneous multiplicity of the living" thus gets rewritten as, once again, the menagerie, the zoo.

This systematic parsing of the animal other into quite different and discrete ontological and ethical categories in turn evinces the obsessive hierarchizing and classification of the other so central to the Enlightenment project, which reaches full flower in the nineteenth and twentieth centuries. And this in turn leads us to the final dimension of *The Garden of Eden* we need to consider. For as Harriet Ritvo, Etienne Balibar, and others have noted, this anxiety about maintaining distinctions and hierarchies in the animal kingdom cannot be separated from a similar anxiety that attends the questions of *racial* and *national* distinctions in post-Enlightenment culture. As Ritvo's painstaking historicist work on Victorian England demonstrates,

> Whether expressed in terms of hybridization or crossbreeding, discussions of animal miscegenation inevitably connected general zoological matters with more narrowly human concerns. Indeed, in a period of global empire and rising nationalism, the zoological and agricultural discussion of these matters—involving mixture and separation, constructed boundaries and carefully analyzed distinctions—may have derived much of its structure, as well as its heated tone, from

its easy compromise of the taxonomic barrier that ostensibly separated
animals from people.[39]

What makes this set of issues so pressing for *The Garden of Eden*, of course,
is not just the prominent species discourse I have been insisting on, but also
the novel's recoding of the transgression of heteronormative codes of sex
and gender as a form of "getting dark" (carried most conspicuously by the
obsessive tanning motif so many critics have noticed), which is in turn
linked to the figure of Africa and, beyond that, to Hemingway's interest in
what he calls "tribal things." As Comley and Scholes have observed, *The
Garden of Eden* is most obviously about how changes in sexuality are publicly
signified by changes in appearance—hair cutting, hair dyeing, clothing
choices, and deep tanning. "What is not so plain in the published book,"
they point out, "but is much clearer in the manuscript is that the darkening
of skin color links this new eroticism to fantasies of miscegenation" (90). As
such, the novel partakes of a rather familiar discursive strategy, in which "the
obsession with tanning is connected with the desire to reach a primal level
of experience, some heart of darkness, from which Euro-Americans have
been cut off by their heritage of enlightenment" (92), so that "the narrative
moves from transgression to transgression, metamorphosis to metamorphosis, closer and closer to Africa" (95).

It probably goes without saying that Hemingway's novel thereby provides what looks like a textbook example of the dynamic described by Bell
Hooks in her well-known essay "Eating the Other," in which otherness "becomes spice, seasoning that can liven up the dull dish that is mainstream
white culture." "The 'real fun,'" she continues, "is to be had by bringing to
the surface all those 'nasty' unconscious fantasies and longings about contact with the Other embedded in the secret (not so secret) deep structure of
white supremacy."[40] For Hooks—and this, I think, has special resonance for
our understanding of Catherine Bourne—"it is precisely that longing for
the pleasure that has led the white west to sustain a romantic fantasy of the
'primitive' and the concrete search for a real primitive paradise, whether
that location be a country or a body, dark continent or dark flesh, perceived
as the perfect embodiment of that possibility" (27). In *Eden*, of course, it is
all of these, as Catherine associates the "devil things" she entices David to
participate in with both the obsessive tanning of her body and the idealized
fantasy site of such transgressions, Africa. "You're my good lovely husband
and my brother too," she tells David. "I love you and when we go to Africa
I'll be your African girl too" (29). "Doesn't it make you excited to have me
getting so dark?" she asks David.

It is this sort of recoding of heteronormative sexual taboo via racist discourse as "getting dark" that leads Toni Morrison in her widely read commentary *Playing in the Dark: Whiteness and the Literary Imagination* to observe that Hemingway's novel reproduces an all-too-familiar "discursive Africanism" of the kind remarked by Hooks.[41] Of Catherine's insistence to David that "I want to be your African girl," Morrison writes: "While we are not sure of exactly what this means to her, we are sure of what Africa means to him. Its availability as a blank, empty space into which he asserts himself, an uncreated void ready, waiting, and offering itself up for his artistic imagination, his work, his fiction, is unmistakable" (88–89).

But what complicates *this* Africa—what we might call "the father's Africa"—is precisely "what Africa means to her," to Catherine, and second, that Africa for nearly the whole novel, even for David, is not at all "a blank, empty space" but rather a *traumatic* site, populated by nonhuman others that he identifies with, a site he has tried for years to avoid confronting, *not* because of its terrifying "primitive" otherness—as in, say, Conrad's *Heart of Darkness*—but rather because it is the place where the animal other is violently subjugated by the code of the white father. In Morrison's reading, as in much of the criticism, what is missing is attention not to Catherine's plight and what it signifies, but rather to the importance of the discourse of species and the ethical problematic of nonhuman others. Morrison is quite right to observe that there are really *two* Africas in the novel: the one of the hunting story, of male bonding and the father-son relationship, "imagined as innocent and under white control," and the "larger Catherine-David Africanist Eden" that "sullies" the first, in which Africanism is "imagined as evil, chaotic, impenetrable" (89). Both are "enabled by the discursive Africanism at the author's disposal" (89–90). But if we take seriously the discourse of species in the novel—and specifically the possibility that the young David's condemnation of the killing of the elephant is in fact critical and not just symptomatic—then we cannot help observing that this first Africa of the father-son relationship is anything but "innocent," a happy site of male bonding. Rather, it is a scene of betrayal, violence, and loneliness that clearly drives David to identify with Catherine's "devilish" rejection of the code of the father and all it stands for.

Moreover, if "Africa" may be said to have a primary association for David as he writes the narrative, it is an association along species lines rather than racial lines, one in which the "darkness" of the place resides not with the land and its "primitive" inhabitants, but rather with the sins of the father. Indeed, David's cross-species identification makes him *more* critically aware of the father's imperialist and patriarchal relation to the discourse of Africanism remarked by Morrison—associations that are repeatedly coupled and condensed during the elephant hunting narrative. "They had found the trail of the old bull finally," we read, "and when it turned off onto a smaller elephant road Juma had looked at David's father and grinned showing his filed teeth and his father had nodded his head. They looked as though they had a dirty secret, just as they had looked when he had found them that night at the shamba" (180). This "dirty secret" doesn't separate the white hunter father from the African guide Juma; rather, it *binds* them in a common sacrificial enterprise that excludes women, animals, and children. And if we fail to understand the significance of species here and the explicit parallel between speciesism, racism, and sexism in this passage, then we make the mistake of improbably equating the "getting dark" of David's cross-identification with the "devil" Catherine with the "getting dark" of the father's and Juma's drunken sexual usage of the *bibis* at the *shamba*.

To put it another way, there is clearly more distance between David and his father here than there is between Juma and the father, who are bound together in patriarchal speciesism *across racial lines*. Thus, when the father nods in acknowledgment when Juma grins, "showing his filed teeth" (which signifies, as one critic has noted, that Juma descends from a cannibalistic tribe), the message we should take from the moment is *not* that Juma is a primitive savage and the father isn't, but rather that the truth of the white father is that he should have filed teeth as well. In fact Juma's cannibalism is, if anything, more honest and rigorous in its refusal to abide by the question-begging sacrificial substitution of animal for human that Enlightenment humanism uses to cover its tracks.

My interest here, then, is similar to what Hooks has in mind when she wonders "whether or not desire for contact with the Other, for connection rooted in the longing for pleasure, can act as a critical intervention challenging and subverting racist domination, inviting and enabling critical resistance" (22). To return to Catherine, then, the point is that "getting dark" for her doesn't *just* signify how her sexual fantasy is structured by Africanist discourse; it *also* signifies her rejection of "getting dark" in the *father's* terms, the *father's* Africa, its violence, its killing of animals, and all that it represents—this is, after all, "what Africa means to her." Thus, the *only* way for Catherine's "getting dark" to be seen as critical and not just stereotypical racist "spice" is to understand that if in one sense it traffics in the discourse reproduced in the father's imperialist Africanism, in another sense—that of identification with the child and the elephant—it unmasks it. In these terms, her fantasies of racial and species miscegenation don't reproduce the father's Africanism because they in fact question its constitutive terms—that is, they question the sacrificial economy (of child, of woman, of animal) that discourse is based on.

In fact, one might well argue that here the ur-mechanism of whiteness's own self-naturalization as a racially unmarked category—its fundamental operation of recoding the always unstable and fluid distinction between white and nonwhite as the more stable and identifiable distinction between human and animal—is subjected to rather devastating exposure and critique. As Etienne Balibar puts it in his essay "Racism and Nationalism," "every theoretical racism draws upon *anthropological universals.*" "In all these universals," he continues,

> we can see the persistent presence of the same "question": that of *the difference between humanity and animality,* the problematic character of which is re-utilized to interpret the conflicts within society and history. . . . Man's animality, animality within and against man—hence the systematic "bestialization" of individuals and racialized human groups—is thus the means specific to theoretical racism for conceptualizing human historicity. A paradoxically static, if not indeed regressive, history, even when offering a stage for the affirmation of the "will" of superior beings.[42]

Following Balibar, we can say that "getting dark" in *The Garden of Eden,* when read in isolation from the novel's species discourse, quite obviously reproduces the Africanist discourse noted by Morrison. But when twinned with the novel's species discourse, it may be seen to unmask that very Africanism because it attacks the racism inherent in that discourse at the fundamental level of "anthropological universals" by exposing the sacrificial economy of speciesism—the unquestioned availability of "animality" as a means of naturalizing and grounding racist discourse—on which racism historically depends. After all, the use of "animality" as a crucial supplement to the discourse of racism is effective only so long as the distinction between human and nonhuman is assumed to be unproblematically coterminous with the distinction between subject and object. Hemingway's novel, then, attacks racism, *but not on the terrain of racial discourse itself,* instead using the "off site" of species discourse to undermine racism's conditions of possibility.

I will end, then, with a transvaluation of the symbolic economy of species and gender, one that takes place both in and beyond the novel itself. In a curious moment in a novel full of curious moments, Colonel Boyle tells Catherine, "I saw you in the Prado looking at the Grecos." "Do you always look at them," he continues, "as though you were the young chief of a warrior tribe who had gotten loose from his councillors and was looking at that marble of Leda and the Swan?" (62). Here Boyle—one of the novel's avatars of the father—finds in Catherine a figure for transformation of gender, race, and species, as she is described as male, as African, and as raptly contemplating a scene of cross-species miscegenation (Comley and Scholes, 95). But here the novel speaks more than Boyle, in the familiar position of white male voyeur, can know. For most readers, under the influence of Yeats's famous version of the story, will imagine a scene of sexual violence, in which a god assumes animal form to overpower his victim. But the only Leda in the Prado is an almost life-size fourth-century sculpture by Timotheos, in which the swan seeks refuge from a pursuing eagle as Leda raises her cloak, in a gesture of cross-species friendship, to hide prey from predator (96).[43] A fitting figure, it seems to me, for how veiled such cross-species relations remain, and how readily we track them down and rewrite them as anything but the wholly other.

CHAPTER 5

Faux Posthumanism

The Discourse of Species and the Neocolonial
Project in Michael Crichton's *Congo*

As we have seen in the foregoing chapters, the philoso-
phy of animal rights, at least in its current state of the art, remains tied to
the theoretical topos of the mirror and the look, and as such it reorients the
question of the alterity of the nonhuman other once again toward the figure of the human. What seems to be needed, then, is a framework for thinking about the problem of subjectivity and species difference in terms of
embodiment and multiplicity rather than identity. That case is made
powerfully—and self-consciously in extremis—by Gilles Deleuze and
Félix Guattari in *A Thousand Plateaus*.[1] Deleuze and Guattari would see the
rights view as firmly circumscribing animal difference within an Oedipal
scenario, one in which all forms of subjectivity must sooner or later be referred for their validation and legitimacy—their legibility, if you will—not
so much to the father but to the name of the father (to invoke the crucial
Lacanian distinction surveyed in chapter 3). In the fascinating sections on
animality in *A Thousand Plateaus*, what Deleuze and Guattari call "becoming-
animal" "always involves a pack, a band, a population, a peopling, in short,
a multiplicity" (239). "We must distinguish between three kinds of animals," they continue:

> First, individuated animals, family pets, sentimental, Oedipal animals
> each with its own petty history, "my" cat, "my" dog. These animals in-
> vite us to regress, draw us into a narcissistic contemplation, and they
> are the only kind of animal psychoanalysis understands, the better to
> discover a daddy, a mommy, a little brother behind them. . . . And then
> there is a second kind: animals with characteristics or attributes;
> genus, classification, or State animals; animals as they are treated in
> the great divine myths. . . . Finally, there are more demonic animals,

pack or affect animals that form a multiplicity, a becoming, a population. (240–41)

Deleuze and Guattari's distinctions aim to underscore that the figure of
the animal, properly understood, is a privileged figure for the problematic
of the subject in the most general sense because here we are forced to confront the reality that the subject is always already multiple.[2] The mistake in
assuming that an animal's ethical standing is to be equated with its singularity, its inhabitation of the space of Identity, is borne out in Deleuze and
Guattari's treatment of Freud's interpretation of the famous case of the
Wolf-Man. "Comparing a sock to a vagina is OK, it's done all the time,"
Deleuze and Guattari tell us, "but you'd have to be insane to compare a pure
aggregate of stitches to a field of vaginas; that's what Freud says." "This represents an important clinical discovery," they continue,

> a whole difference in style between neurosis and psychosis. For ex-
> ample, Salvador Dali, in attempting to reproduce his delusions, may
> go on at length about THE rhinoceros horn; he has not for all of that
> left neurotic discourse behind. But when he starts comparing goose-
> bumps to a field of tiny rhinoceros horns, we get the feeling that the
> atmosphere has changed and that we are now in the presence of mad-
> ness. Is it still a question of a comparison at all? It is, rather, a pure
> multiplicity that changes elements, or *becomes*. On the micrological
> level, the little bumps "become" horns, and the horns, little penises.
>
> No sooner does Freud discover the greatest art of the unconscious,
> this art of molecular multiplicities, than we find him tirelessly at work
> bringing back molar unities, reverting to his familiar themes of *the* fa-
> ther, *the* penis, *the* vagina, Castration with a capital *C*. (27)

It is *Freud*, then (and for Deleuze and Guattari psychoanalysis as such), who
is engaged in repression—in this case, repression of the "important clinical
discovery" that the unconscious is first and foremost a power of multiplicity and becoming, one whose truth the bourgeois, patriarchal Freud must
disavow by misreading the Wolf-Man's psychosis as mere neurosis. "The reductive procedure of the 1915 article is quite interesting," they continue;
Freud holds that "the comparisons and identifications of the neurotic are
guided by representations of things, whereas all the psychotic has left are
representations of words. . . . Thus, when there is no unity in the thing,
there is at least unity and identity in the word" (27–28). But Freud's patient,
who inhabits the psychotic universe of multiplicity, knows better. He
knows, as Deleuze and Guattari put it, that

the only thing Freud understood was what a dog is, and a dog's tail. It wasn't enough. It wouldn't be enough. . . . [The Wolf-Man] knew that he was in the process of acquiring a veritable proper name, the Wolf-Man, a name more properly his than his own, since it attained the highest degree of singularity in the instantaneous apprehension of a generic multiplicity: wolves. He knew that this new and true proper name would be disfigured and misspelled, retranscribed as a patronymic. (26–27)

As we are about to see, the retranscription of the becoming and multiplicity of the nonhuman other in the form of the patronymic turns out to be an especially effective strategy for tethering the category of subjectivity to the neocolonial project, even when—*especially* when—the transcriber is a nonhuman animal.

. . .

The relation of language, identity, and species is at center stage in Michael Crichton's novel *Congo*, originally published in 1980 and reissued nearly a decade and a half later in conjunction with the box-office flop of the same name.[3] Crichton's novel is a beguiling jumble of factoids assembled within the frame of "an old-fashioned thriller-diller" (as one of the jacket blurbs puts it), and it is made all the more inscrutable by its affective flatness, its characteristic postmodern depthlessness (if we believe Fredric Jameson's description),[4] which gives us a novel made up of little *other* than plot and information, a novel with precious little time for the psychological depth of character usually associated with the novel in its earlier forms. But I am less interested in Crichton's text on aesthetic grounds than in how it exemplifies the discourse of species within postmodern culture and, within that, the moment of neocolonialism. Crichton's novel seems to provide a resolutely "progressive" engagement—and this squarely within the mainstream of American mass culture—with the ethical and ethological question of the animal. Despite the title and the novel's heavy debt to the paradigm established for modernism by Joseph Conrad's *Heart of Darkness* (and later for American mass culture by Francis Ford Coppola's *Apocalypse Now*), *Congo*—in its staging of nonhuman subjectivity—immediately promises something different.

That promise is carried largely if not solely by the central character of the novel, a mountain gorilla named Amy who has been raised by Dr. Peter Elliot in a language lab at the University of California at Berkeley. Amy has prodigious linguistic abilities, even beyond those of real-world apes like Koko and Washoe; she has a 620-item vocabulary in Ameslan sign language and even (like the real-life bonobo Kanzi at Georgia State University)[5] understands some spoken English. As the plot unfolds, Amy has been having dreams and making finger paintings of what is later revealed to be the Lost City of Zinj (which she remembers from her infancy). She accompanies Elliot and an expedition from Earth Resources Technology Services (ERTS), which has been funding Elliot's research, to Zaire to search for rare superconductive "blue" diamonds, which are particularly useful for future post-silicon-chip information technologies. Led by Dr. Karen Ross, a ruthlessly competitive and analytical twenty-three-year-old mathematical whiz who thus far has merely supervised field parties by satellite link from home base in Houston, the expedition must succeed where a previous one failed. For as the book opens, we discover that the first ERTS party sent to look for the diamonds has been violently murdered, their skulls mysteriously crushed with a force surpassing that of the strongest human.

And so—skirting cannibals and political unrest and the machinations of a competing expedition from a Euro-Japanese consortium—Ross's group, led by an unscrupulous but essentially honorable former great white hunter and mercenary named Munro, makes its way into the deepest rain forest of the Congo. They discover that the mythical "Lost City of Zinj" does indeed exist, and that an unimaginably rich lode of blue diamonds is indeed deposited there, at the foot of the volcanic Mount Mukenko. What they *also* discover, however, is that the entire area is patrolled by a previously unknown species of gray gorilla that, as Elliot observes, has "been single-mindedly bred to be the primate equivalent of Doberman pinschers—guard animals, attack animals, trained for cunning and viciousness" (252). These creatures have guarded the blue diamond mines at Zinj for five hundred years, handing down their own culture and behavior—and most important, "a language system far more sophisticated than the pure sign language of laboratory apes in the twentieth century," one that combines a "wheezing" type of vocalization with a gestural repertoire "rather like Thai dancers" (258). The gray gorillas also can make and use stone tools—specifically the crescent-shaped stone paddles that they use as weapons against all intruders. It is these creatures, of course, who are responsible for the gruesome fate of the first ERTS expedition. And they threaten to wipe out the second group until fellow gorilla Amy translates enough of their vocabulary to enable the expedition to broadcast into the jungle a set of simple messages recorded by Amy—GO AWAY, NO COME, BAD HERE—that makes the gorillas halt their final, highly coordinated assault just in the nick of time

(276–81). As the novel ends, the threat of the gray gorillas is removed once and for all as Mount Mukenko suffers a massive eruption and the ERTS expedition escapes a final attack—this time by the cannibalistic Kigani tribesmen—through the deus ex machina of a hot-air balloon left behind at the site where a plane crashed with the ill-fated Euro-Japanese consortium.

As the heavily freighted literary and cultural inheritance of the novel's title more than suggests, it is nearly impossible *not* to read Crichton's novel as a kind of racial allegory that uses the discourse of species to recode deeply held fantasies of racial identity that were alive and well at the dawn of the Reagan era, when the novel was published. From this vantage, the novel may be seen as firmly circumscribed within the discourse of Africanism identified by Toni Morrison in the previous chapter, "deployed as rawness and savagery" and crucial to the white American literary imagination by serving for it "duties of exorcism and reification and mirroring."[6] Read in this way, the novel provides a cautionary tale to white, technocratic, upwardly mobile America in the early Reagan years about the dangers of believing that "blackness" can be domesticated and made productive for the social project. In this light, the moral of the gray gorillas and their rebellion against their masters would be this: Even if you "whiten" them up a little from black to gray with language and learning, in the end they will only use it to rebel against you. Like their twins the cannibalistic Kigani, they will kill you the first chance they get, so better to leave them in deepest, darkest Africa. In other words, *Congo* is a vintage cultural document of early 1980s laissez-faire. And the eruption of Mount Mukenko at the novel's end emphatically punctuates the point: No matter how good your technology and your intentions, it is better to understand that blackness is an-other country, which is why even Amy finally can't be domesticated. In the end she too must return to the jungle whence she came, because blood is thicker than culture.

It is no doubt useful—and in a longer treatment would be imperative—to read Crichton's novel as an allegory of racial fantasy in the United States of the early eighties. But I want to focus instead on how the discourse of species serves to organize and enable the novel and its ideological project. After all, what is much more remarkable and unusual than the racial allegory in the novel is that it doesn't just entertain but in fact turns on the questions of nonhuman subjectivity, intelligence, language, and culture. And it makes a point of ballasting its systematic questioning of speciesist assumptions about all of these categories with references (both real and imagined) to the literature on ape-language experiments (32–33), field studies of animal societies (178) and their tool use (250), and human beings' more general "complacent egotism with regard to other animals" (253).[7] Elliot's musings in the following passage are in the dominant key of the novel's handling of the problem of nonhuman subjectivity:

> Over the years, he had come to feel acutely the prejudices that human beings showed toward apes, considering chimpanzees to be cute children, orangs to be wise old men, and gorillas to be hulking, dangerous brutes. . . .
>
> Elliot had witnessed repeatedly the human prejudice against gorillas, and had come to recognize its effect on Amy. Amy could not help the fact that she was huge and black and heavy-browed and squash-faced. Behind the face people considered so repulsive was an intelligent and sensitive consciousness, sympathetic to the people around her. It pained her when people ran away, or screamed in fear, or made cruel remarks. (113)

Crichton's novel is chock-full of passages like this, and it thus immediately confronts us with the question of how we are to relate this apparent de-centering of the figure of the human to that *other* central feature of the novel's universe: an immense technoscientific apparatus driven to dizzying accomplishments (so the drift of the novel goes) under the spur of free market global capitalism, all seeming to immediately *re*center the figure of the human via the utterly conventional privileging of the tool-using, techno-logical capacities of *Homo sapiens*.

The most direct way to address this dilemma is to understand—as we will see below—that as the novel unfolds, each half of the constitutive dichotomy of the discourse of speciesism ("human/animal") undergoes a systematic bifurcation: between Amy and the gray gorillas on the "animal" side, and between the ERTS party and their primitive "others," the cannibalistic Kigani, on the side of the "human." The novel will then reconstitute these elements not along the lines of biological species, but rather in terms of a double articulation: first, according to the logic of the *discourse* of species, and second, according to the place of each character or group in terms of its serviceability to the imperatives of neocolonialism.

As for the bifurcation within the category of animality itself, it is governed by terms very close to those deployed by Deleuze and Guattari in *A Thousand Plateaus*. Amy, as the homophonic echoes of her name suggest (e.g., "a me," "hey, me," as if to remind us hypogrammatically, "hey, it's me I see when I look at her"), is thoroughly inscribed within the singular, individuated, and finally Oedipalized regime of subjectivity. She is clearly a diminished form of the human, a "narcissistic" reflection who has something very close to the status of a pet for Elliot. Like a good Freudian subject who evinces "the cultural trend toward cleanliness" that "originates in an urge to get rid of the excreta,"[8] Amy finds "bodily excretions suitable terms to express denigration and anger," and more than once when she is angry she signs *"Peter shitty"* (174). She loves to be tickled, enjoys the occasional cigarette, and rejects jungle bananas because they are slightly sour (she prefers milk and cookies) (231). And when she wanders off from the expedition into the rain forest, she tells Elliot that she left because she was jealous of Karen Ross (*"Peter like woman no like Amy"* [229])—an Oedipal triangulation reinforced in a conversation between the guide Munro (who, "Instead of patting her on the head and treating her like a child, as most people did," "instinctively treated her like a female" [152]) and a group of pygmies, which Munro recounts to Elliot:

> They wanted to know if the gorilla was yours, and I said yes. They wanted to know if the gorilla was female, and I said yes. They wanted to know if you had relations with the gorilla: I said no. They said that was good, that you should not become too attached to the gorilla, because that would cause you pain.
>
> Why pain?
>
> They said when the gorilla grows up, she will either run away into the forest and break your heart or kill you. (166–67)

We can add to this Oedipalizing Amy's privileging of linguistic ability. She refers to normal forest gorillas as *"dumb"* because they *"no talk"* (230), and in this, she is like the languaging chimpanzee referred to earlier in the book, who calls nonlanguaging chimps "black things" and who, when asked to sort photographs of chimps and humans, "sorted them correctly except that both times he put his own picture in the stack with the people" (45). And later in the story, Amy calls the gray gorillas *"dumb things"* (267) because they fall for her masquerade as Elliot's mother—a ploy that narrowly saves his life when he falls down a slope and finds himself surrounded by the strange and dangerous animals. Most telling, perhaps, in fixing Amy's status as a diminished form of the human, is that she dreams, but her dreams need no interpretation; they turn out to be iconically transparent representations of the jungle home she remembers from her childhood, not manifest symbolic transformations of a latent dream text read only through interpretive work. Like us, Amy dreams; unlike ours, her dreams are simple.[9]

The Freudian dimension of the novel's discourse of species is symbolically mapped quite well at two key moments early on. As the novel opens, we witness the character Kruger, a member of the ill-fated first ERTS expedition, preparing for the daily video link to Houston, musing as he works about "the way Americans always put on a fresh shirt and combed their hair before stepping in front of the camera. Just like television reporters." Then suddenly,

> Something struck him lightly in the chest . . . a fleshy bit of red fruit rolled down his shirt to the muddy ground. The damned monkeys were throwing berries. He bent over to pick it up. And then he realized that it was not a piece of fruit at all. It was a human eyeball, crushed and slippery in his fingers, pinkish white with a shred of white optic nerve still attached at the back. (5)

He looks for his companion Misulu, who has suddenly vanished, "And then he heard the wheezing sound again" (5). Moments later, after discovering the body of Misulu with its skull crushed, Kruger himself is attacked by the gray gorillas.

This passage establishes from the outset the discursive coordinates of speciesism that I have been discussing thus far. For what strikes Kruger in the chest is nothing other than the privileged sensory apparatus of the Freudian "human" as it is forcibly ejected from Misulu's skull when his head is crushed by the stone paddles the gray gorillas use in their attacks. And the Freudian eye is doubled here by the eye of the video camera, before which humans display species-specific cultural behaviors (the fresh shirt, the combed hair) that performatively define them *as* "human" as they preen in aesthetic contemplation. All of this is well glossed by Olivier Richon's observation that in the Freudian scheme "the aesthetic runs counter to the instinctual. The aesthetic involves vision and therefore separation. The aesthetic, unlike the instinctual, erects a barrier between species; it puts emphasis upon the object of desire, rather than desire itself."[10]

The Freudian scheme is only confirmed by the savage attack of the gray gorillas, who (as befits their "animalistic" status) reduce humanism's privileged sensory organ to a mere glob of tissue easily mistaken for vegetable matter thrown by disdainful (and dung-throwing) monkeys, all punctuated, in effect, by Kruger's discovery. The difference between the Freudian symbolic eye and its rewriting—perhaps we should say *un*writing—by the gray gorillas is further reinforced by the gorillas' most epithetical attribute throughout the novel: their "soft wheezing," which Kruger at first hearing mistakes for a big cat with "respiratory trouble." This in turn secures ever more firmly the association of the gray gorillas with the domain of what Slavoj Žižek, following Lacan (and beyond that, Kant) calls "the Thing," *das Ding*, "'the flesh from which everything exudes,' the life substance in its mucous palpitation" that is literalized in Kruger's initial misperception of the sound as gurgling mucus. "The very notion of life," Žižek reminds us, "is alien to the symbolic order"—and, need we add, to its privileged expression in this novel, technoscience.[11]

As with my discussion of Hemingway in the previous chapter, however, my invocation of Žižek and Lacan should not be taken to imply that an anatomy of the discourse of species in *Congo* depends on a psychoanalytic reading alone. Indeed, the stridently antipsychoanalytic analysis of animality by Deleuze and Guattari in *A Thousand Plateaus* (or perhaps one should say postpsychoanalytic, since their critique is indeed indebted to the insights of Lacan) provides an equally powerful tool for laying bare the discursive work of species in the novel. For them, as I have already noted, the category of the animal is read less in terms of its status as the traumatic "Thing" and more in light of "becoming" and "multiplicity"—a reading that extends to "schizophrenic" extremes the well-known critique in Max Horkheimer and Theodor Adorno's *Dialectic of Enlightenment* of the domination of nature by the Enlightenment subject (and its apotheosis in Enlightenment science—very much to the point here). As Adorno writes in a passage from *Aesthetic Theory* that condenses many of the key themes of the work with Horkheimer, "Nature, whose imago art aspires to be, does not yet exist; what is true in art is a non-existent. It comes to coincide with art within that Other, which a reason fixated on identities and bent on reducing it to sheer materiality calls Nature. *That other is, however, neither a unity nor a single concept, but rather the multiple.*"[12] It is that "multiple" that Deleuze and Guattari aim to unleash in their attempt to move not only beyond the Enlightenment that renders the other as an undifferentiated mass whose name is *das Ding* or "Nature," but also beyond the dialectic (as Adorno himself strained to do in his "negative dialectics"). As they put it in *A Thousand Plateaus*—and here again there is a strong echo of Horkheimer and Adorno—the concept of multiplicity "was created precisely in order to escape the abstract opposition between the multiple and the one, to escape dialectics, to succeed in conceiving the multiple in the pure state, to cease treating it as a numerical fragment of a lost Unity or Totality or as the organic element of a Unity or Totality yet to come" (32).

The concept of the animal as "multiplicity" is crucial to understanding that in Crichton's novel the gray gorillas represent for Enlightenment subjectivity not only a monstrous return of the repressed ("the Thing" as glossed by Lacan and Žižek), but also a more specifically Deleuzean-Guattarian disturbance of the symbolic field, one that cannot be *located* and *fixed* as the negative moment or pure exteriority of an identity term (whether psychoanalytic or dialectical). The novel prepares us for such a

reading early on, when the image of one of the gray gorillas responsible for
the massacre of the first ERTS party cannot quite be made out on the Houston tape of the video feed. At first we are told, "They could see the outline
of the shadow now. It was a man" (13); then we get reservations: "It did not
look to her like a limping man; something was wrong. She couldn't put her
finger on what it was" (14). Finally, under the pressure of "data salvage" by
complex computer image-enhancement processes, "the image 'popped,'
coming up bright and clean. . . . Frozen on the screen was the face of a male
gorilla" (20). Here we see the power of those prosthetic extensions of the eye
glossed earlier in this book by Haraway, those imaging technologies—like
the satellite and computer networks sustaining the expedition and mapping
the planet—associated with the "unregulated gluttony" of phallic vision. In
them "all perspective gives way to infinitely mobile vision, which no longer
seems just mythically about the god-trick of seeing everything from
nowhere, but to have put the myth into ordinary practice."[13]

But no sooner does the singular face of the gorilla emerge from the computer image field than it threatens to vanish into the buzzing forest of information space: "In the highly sophisticated data-processing world of
ERTS, there was a constant danger that extracted information would begin
to 'float'—that the images would cut loose from reality, like a ship cut loose
from its moorings. This was true particularly when the database was put
through multiple manipulations—when you were rotating 10^6 pixels in
computer-generated hyperspace" (23–24). And even when ERTS "buys"
the image as a picture of a gorilla, its meaning still cannot be nailed down—
not even when primatologist Elliot comes on board:

> Elliot was not so sure. He reran the last three seconds of videotape a
> final time, staring at the gorilla head. The image was fleeting, leaving
> a ghostly trail, but something was wrong with it. He couldn't quite
> identify what. . . .
>   Elliot was sure this creature was too light to be a mountain gorilla.
> Either way they were seeing a new race of animal, or *a new species*. (75)

This disquieting nonidentity of the gray gorillas erupts later in the novel
into full-blown "demonic" Deleuzean-Guattarian multiplicity as the second ERTS expedition settles into camp near the Lost City. During the
penultimate battle, "The gorillas attacked from all directions; six of them
simultaneously hit the fence and were repelled. . . . Still more charged,
throwing themselves on the flimsy perimeter mesh. . . . And then he saw gorillas in the trees overhanging the campsite. . . . Elliot turned and saw more
gorillas tearing at the fence" (241). The seething, teeming multiplicity of

the gray gorilla troop is registered even more forcefully as the expedition tries to plot a line of flight the next day. As they move into the forest to scout a route, "Munro was disturbed by what he saw; some trees held twenty or thirty nests, suggesting a large population of animals." Then "he looked off and 'had the shock of my bloody life. Up the slope was another group, perhaps ten or twelve animals—and then I saw another group—and another—and another still. There must have been three hundred or more. The hillside was *crawling* with gray gorillas'" (258). Munro's vision marks the animals as even more "demonic" in their insectlike multiplicity, their "population" (to use Delueze and Guattari's terms), when we are immediately told, in a vintage Crichton factoid, that "the largest gorilla group ever sighted in the wild had been thirty-one individuals, in Kabara in 1971" (258). The expedition tries to plot another route out of the forest, away from the Lost City, but it's no use: "He checked his watch: they had been gone ten minutes. And then he heard the sighing sound. It seemed to come from all directions. He saw the foliage moving before him, shifting as if blown by a wind. Only there was no wind. He heard the sighing grow louder" (261).

Here and elsewhere, it is clear that the gray gorillas represent not the individualized and Oedipalized half of the bifurcated category of animality inhabited by Amy, but rather the demonic multiplicity of Deleuze and Guattari's "pack" animality. From this vantage, it is a delicious Deleuzean irony perfectly befitting the symbolic significance of the gray gorillas that Elliot, in the end, will be denied his one obsession: "He found he was bored by the prospect of further exploration of Zinj; he had no interest in diamonds, or Amy's dreams; he wished only to return home with a skeleton of the new ape, which would astonish colleagues around the world" (247). But it is not to be; the multiplicity of the gray gorillas is not to be reduced to the patronymic of the Latin scientific nomenclature—the singular specimen, the representative example—for the eruption of Mukenko buries the unknown species in ash, wiping them out entirely.

· · ·

The central problem for the novel, then, is how this "multiplicity" of the animal other will be managed, a problem made all the more acute within a discursive context that has already, at least on the face of it—through the character of Amy and through the various ethological factoids—called into question the traditional containment strategies of the Enlightenment discourse of animality. Here again the Frankfurt school can be of use, particularly in light of Michael Taussig's highly original rereading of their work on the problem of mimesis. For once the familiar Enlightenment and Freudian categories of speciesism are destabilized, as they surely are in this text, the problem we are immediately confronted with is one that we saw animating the "confusions" of *The Silence of the Lambs*—what Philippe Lacoue-Labarthe calls the "primal status and undivided rule of mimetic confusion,"[14] a problem readily thematized by the cross-species presence of language and culture throughout *Congo*. Who or what is miming, and who or what is mimed? Who is "like" and who is "same," who the original and who the copy, who the human and who the animal?

This confusion—which for Lacoue-Labarthe is, strictly speaking, a product of the inescapability of representation as such from Greek civilization to the present—is of particular moment in the context of Taussig's reading of Walter Benjamin's analysis of mimesis. "Nature creates similarities," Benjamin writes in "On the Mimetic Faculty": "One need only think of mimicry. The highest capacity for producing similarities, however, is man's. His gift of seeing resemblances is nothing other than a rudiment of the powerful compulsion in former times to become and behave like something else. Perhaps there is none of his higher functions in which his mimetic faculty does not play a decisive role.[15] What this means, as Taussig puts it, is that the mimetic faculty is "the nature that culture uses to create second nature" (xiii). "The ability to mime, and mime well, in other words, is the capacity to Other" (19). For Taussig as for Lacoue-Labarthe, the mimetic is thus *always* a site for potential confusion of like and same.

But for Taussig, the mimetic also has a special historical status specific to the era of postcolonialism. Mimetic confusion—or what Taussig calls "mimetic excess"—is generated at a historically unprecedented level when, for example, Cuna women integrate "mousetraps, lunar modules, and baseball games into the traditional scheme of their appliqued shirtfronts—the famous *molas*, international sign of Cuna identity" (132). When this "mimesis of mimesis" becomes globally generalized, we suddenly find ourselves, Taussig argues, in a dizzying but potentially liberating state of affairs:

> History would seem to now allow for an appreciation of mimesis as an end in itself that takes one into the magical power of the signifier to act as if it were indeed the real, to live in a different way with the understanding that artifice is indeed natural, no less than nature is historicized. Mimetic excess as a form of human capacity potentiated by post-coloniality provides a welcome opportunity to live subjunctively as neither subject nor object of history but as both, at one and the same time. (255)

What Taussig characterizes as the reversibility of power relations made possible by mimetic excess is also a central concern of Homi Bhabha's work on what he calls the "ambivalence" of "colonial mimicry," in which the colonizing power attempts to produce the colonized as a little imitation of itself, even while a specifically "native" content must be sustained in the negotiation, if only to justify the "civilizing" work of the colonizer. What we find in colonial mimicry, Bhabha writes, is "the desire for a reformed, recognizable Other, *as a subject of a difference that is almost the same, but not quite.* Which is to say, that the discourse of mimicry is constructed around an *ambivalence;* in order to be effective, mimicry must continually produce its slippage, its excess, its difference."[16]

Colonial mimicry is "therefore stricken by an indeterminacy" (86), and as in Taussig's reading, this indeterminacy cuts both ways. On the one hand, it "fixes the colonial subject as a 'partial' presence," "incomplete" and "virtual" (86), a *"metonymy of presence"* (89) whose form is "almost the same, but not quite." On the other hand, the colonized, by engaging in what Taussig would call the "mimesis of mimesis," can "make the signifiers of authority enigmatic in a way that is 'less than one and double.' They change their conditions of recognition while maintaining their visibility; they introduce a lack that is then represented as a doubling of mimicry. This mode of discursive disturbance," Bhabha continues, "is a sharp practice, rather like that of the perfidious barbers in the bazaars of Bombay who do not mug their customers with the blunt Lacanian *vel,* 'Your money or your life,' leaving them with nothing. No, these wily oriental thieves, with far greater skill, pick their clients' pockets and cry out, 'How the master's face shines!' and then, in a whisper, 'But he's lost his mettle!'" (119).

In this way the colonized sets up "another specifically colonial space of the negotiations of cultural authority" (119). And this means that the site of the production of colonial discourse is "a space of *separation*—less than one and double—which has been systematically denied by both colonialists and nationalists who have sought authority in the authenticity of 'origins.' It is precisely as a separation from origins and essences that this colonial space is constructed" (120).

Taken together with Taussig's reading, Bhabha's analysis helps to underscore both the precise character of this space inhabited and reproduced by Crichton's novel and at the same time the potential power of mimetic excess and reversibility within that space. What is especially striking about *Congo* in this light, however, is that mimetic excess and all it signifies in these postcolonial critiques is never really a threat to the novel's ideological project. Or more precisely, there *is* mimetic confusion aplenty in *Congo,* but the novel deploys an especially effective strategy for managing it—and in so doing, for managing the possible eruption of the "multiplicity" that the animal other signifies. Here is where the discourse of species is crucial to the novel's efforts at ideological recontainment, and why that discourse must therefore be viewed in *Congo* not merely as a counter for questions of race or nation but as irreducible. For in Crichton's novel, mimetic confusion is strategically quarantined within the category of the animal *only*, as a problem to be negotiated between the First World gorilla Amy and the Third World gray gorillas. In fact, one might even say that in this book mimetic confusion is a problem *for* animals only. Here mimetic excess and the potentially liberating confusion it generates do not open onto an interrogation of the category of the human (and within that, the colonizer) but get strategically rewritten as the question, How do you know a *real* gorilla when you see one?

The novel therefore seems—but only seems—to radically question the discourse of speciesism, while at the same time it leaves intact the category of the human and its privileged forms of accomplishment and representation in the novel: technoscience and neocolonialism. Indeed, all of these are simply presented as the more or less "natural" outcome of an evolutionary process governing both nature and geopolitics. As Munro puts it at one point—in a passage that could serve as the novel's credo: "The purpose of life is to stay alive. Watch any animal in nature—all it tries to do is stay alive. It doesn't care about beliefs or philosophy. Whenever any animal's behavior puts it out of touch with the realities of its existence, it becomes extinct. The Kigani haven't seen that times have changed and their beliefs don't work. And they're going to be extinct" (150). Munro understands the law of the jungle—which is to say, in the ideological space of *Congo*, he understands the law of second nature, of capitalism's global market.

One of the more interesting negotiations of the threat of mimetic confusion in the novel occurs when the ERTS party discovers in the Lost City a massive statue of a gorilla with arms outstretched, paddles in hand (237). They think at first that they have discovered an ancient "cult of the gorilla," replete with a priestly caste, and Ross offers "an elaborate explanation" (248) of a culture that might be based on this totemic cross-species identification. As it turns out, though, they have it "all backward." What they have discovered (as frescoes in the building confirm) is a training facility where the gray gorillas are raised and drilled as ruthless guards of the diamond mines. This process depends, of course, on the highly developed mimetic faculties of the apes themselves, but what is most important here is how the prospect of cross-species mimetic confusion, and the possible disruption of the hierarchical relation between human and animal it might hold, is immediately foreclosed and overwritten by what we discover is in essence a relationship of master and slave—a relationship that doesn't just preserve the old hierarchy but intensifies it.

Even more striking is how the most important single instance of mimesis in the book—the taped playback of Amy's wheezing imitation of the gray gorillas' language (GO AWAY, NO COME, BAD HERE) that makes the gorillas halt their attack—takes with one hand from the category of the animal what it appears to give with the other. The significance of this moment is perhaps best understood in terms of its rather uncanny reenactment of the relation between human, machine, and animal recounted in Taussig's analysis of the famous RCA Victor logo of the dog Nipper listening to the phonograph, titled "His Master's Voice." As Taussig puts it, "the power of this world-class logo lies in the way it exploits the alleged primitivism of the mimetic faculty. Everything, of course, turns on the double meaning of fidelity (being *accurate* and being *loyal*), and on what is considered to be a mimetically astute being—in this case not Darwin's Fuegians but a dog" (213). Moreover, Taussig, continues, "there is the curious mimetic gestus of the dog, its body as well as its face miming the human notion of quizzicality. This dog is testing for fidelity and is also a little mystified. What could be more 'human' (or at least anthropomorphic) than this . . . ? *Where politics most directly enters is in the image's attempt to combine fidelity of mimetic reproduction with fidelity to His Master's Voice*" (223; emphasis mine).

From this vantage, the full force of the novel's climactic mimetic moment may now be felt:

> He saw the gorilla bearing down on him. He tensed his body. Six feet away, the charging gorilla stopped so abruptly that he literally skidded in the mud and fell backward. He sat there surprised, cocking his head, listening. . . .
>
> Elliot saw another gorilla stop to listen—then another—and another—and another. The compound took on the quality of a frozen tableau, as the gorillas stood silent in the mist.
>
> They were listening to the broadcast sounds. (280)

Like RCA's dog in Taussig's reading, the gray gorillas are here reinscribed under the sign of "fidelity," whose ambiguity is, as Taussig takes pains to emphasize, entirely to the political point. In this "frozen tableau," "fidelity" names not only the gray gorillas' powerful mimetic faculties, and not only Amy's accomplished mimesis of that mimesis that now seizes and redirects the flow of mimetic power, but also the gray gorillas' trainability, their automatonlike

loyalty to the command issued by "His Master's Voice." In thus circumscribing and redirecting the mimetic situation, what we might call "the name of the
trainer" recontains the potentially threatening multiplicity of the gray gorilla
band under the sign of the human and Oedipal patronymic. As in Freud's interpretation of the case of the Wolf-Man, the word, the signifier, "the semiotic capital *S*" makes multiplicity yield to identity, becoming to being, body
to law. As Deleuze and Guattari put it, "Freud counted on the word to reestab-
lish a unity no longer found in things. Are we not witnessing the first stirrings
of a subsequent adventure, that of *the* Signifier, the devious despotic agency
that substitutes itself for asignifying proper names and replaces multiplicities
with the dismal unity of an object declared lost?" (28).

Such is the resonance, I think, of the "humanizing" elements of the
"frozen tableau," the gorillas comically falling on their rumps in the mud,
cocking their heads quizically like the pet dog in RCA's logo. Here too the animal other is accorded impressive mimetic prowess, only to have it immediately put to the service of a mechanical obedience whose most famous name
in the philosophical tradition, as we know from earlier chapters, is Descartes.
The point here is not so much that the gray gorillas are easily duped, but
rather that the mimetic excess that might destabilize the various dichotomies
I have been discussing is recontained. Mimetic excess, as Taussig puts it, lib-
erates the original from its authoritative status as origin, "drawing attention
to the exuberance with which it permits the freedom to live reality as really
made up" and thus providing "a welcome opportunity to live subjunctively
as neither subject nor object of history but as both, at one and the same time"
(255). But in *this* instance, mimetic prowess is immediately recontained as
obedience, and the trajectory of mimetic production—so ambivalent at the
site of the colonial, as Bhabha points out—is seized by the First World primates (both ape *and* human): a reversal aptly symbolized, as Taussig points
out, by the "blossoming ear-trumpet of the phonograph" (223) of the RCA
logo that mimics the organ of mimetic *reception* but in fact carefully orches-
trates mimetic production in the services of "fidelity"-as-loyalty.

As the novel moves toward its climax, Elliot speculates, in a moment apparently unaware of its own irony:

> To them we are just animals, he thought. An alien species, for which
> they have no feeling. We are just pests to be eliminated.
>   These gorillas did not care why human beings were there, or what
> reasons had brought them to the Congo. They were not killing for
> food, or defense, or protection of their young. They were killing be-
> cause they were trained to kill. (279)

What is so striking and so ideologically symptomatic about *Congo*—and part of what marks it as a mainstream American cultural product of the 1980s— is that it makes no connection *at all* between sentiments like these and the institution of neocolonialism. And it is here, I think, that we can locate the discourse of species in the novel within the larger project of neocolonialism identified by Bhabha and Taussig. For when we try to sort out the potential mimetic confusion broached as nowhere else in the novel in the relation of Amy to the gray gorillas, what we discover is that Crichton's text manages potential mimetic confusion in relation to species distinctions through what Taussig, following Horkheimer and Adorno, calls that "essential component of socialization and discipline," the "organized control of mimesis" (219, 215). In this light, Amy's mimetic abilities are to be distinguished from those of the gray gorillas insofar as they have become Westernized (that is to say, "socialized"), insofar as they take their place as simply another circuit in what Haraway calls the "Command-Control Communications and Intelligence" (C3I) network of mimetic control at the disposal of the expedition and, beyond that, multinationals like ERTS. In fact—as my references to Deleuze and Guattari suggest—language as such in the novel is not a mechanism for the dissemination of difference but is instead chiefly a "patronymic" mimetic technology that must be distributed (to apes, for instance) and maximized while managing the risk of mimetic confusion and reversibility that accompanies it.

From this vantage, it is clear (as Bhabha's analysis helps underscore) that the central question in the novel is not who has mimetic ability and who doesn't, but rather who has mimetic *power* and who doesn't, whose mimetic abilities can control the directionality of symbolic, economic, and political reproduction in the services of the neocolonial project, whose have undergone not repression, exactly (which would mark the text as modernist in either a Jamesonian or a Žižekian frame), but rather maximization and *management*. Crichton's neocolonial Congo thus continues to be a "heart of darkness" that offers up its proverbial cautionary tale about the limits of the "organization of mimesis"—but with a difference, as Bhabha might say. For those limits turn out to be not about the traumatic encounter with the primitive that threatens to activate the animal in all of us (as in modernism proper) but instead are recast as "postideological" problems of "research and development" that are internal to a more properly postmodern and neo-colonial world system. In this system, as Haraway puts it, "the reproduction of capitalist social relations" depends on "an engineering science of automated technological devices, in which the model of scientific intervention is technical and 'systematic,' . . . [t]he nature of analysis is technological functionalism, and ideological appeals are to alleviation of stress and other signs of human obsolescence."[17]

Nowhere is this borne out more clearly than in the figure of Amy herself, who is linked in nearly cyborgian fashion to the tape recorders, VTRs, satellite systems, and software banks back in Houston in the group effort to decode the language of the gray gorillas. And Amy has her place within the reproduction of capitalist social relations as well. Elliot speculates about Amy's visit to the Congo as the first test of "the Pearl thesis," in which "we can imagine language-skilled primates acting as interpreters or perhaps even ambassadors for mankind, in contact with wild creatures" (65). Such a notion may seem nearly revolutionary in its imagining of a proverbial "first encounter" between First World and Third World subjects who are not human, but Elliot has his doubts, because as one critic of the thesis points out in the novel, at places like the language labs of Berkeley "we are producing an educated animal elite which demonstrates [toward its wild counterparts] the same snobbish aloofness that a Ph.D. shows toward a truck driver" (66). Here the "capitalist social relations" remarked by Haraway are reproduced *within the category of animality itself,* and the point, of course, is that Amy, an upwardly mobile professional who deals in symbolic knowledge (as Robert Reich or Alvin Toffler might say), has much more in common with fellow educated yuppies Ross and Elliot than with any wild ape—as she herself, we presume, would be the first to point out.

What this suggests, then, is that the discourse of species in *Congo* is rearticulated on the more fundamental ur-discourse of the "organization of mimesis" by the world system of global capitalism in its neocolonial moment. In this system, as Jameson puts it, decolonization goes hand in with neocolonialism and we find, "symbolically, something like the replacement of the British Empire by the International Monetary Fund,"[18] a process that uses third wave technoscience to extend its knowledge-for-profit of the colonized other (whether "nature," "primitive" civilizations, indigenous peoples, or native animals) to ever more capillary levels. The central ideological symptom of Crichton's novel, then, is the jarring disjunction between its seemingly progressive discourse of species (in which we are apparently disabused of many of our speciesist attitudes about nonhuman others) and its thorough taking for granted of neocolonialism, in which one senses nostalgia for the good old days of outright imperialism, when "Nairobi was a fast-living place indeed." "The men were hard-drinking and rough, the women beautiful and loose, and the pattern of life no more predictable than the fox hunts that ranged over the rugged countryside each weekend" (109). Thank god, we are to surmise, at least the guards at the Nairobi airport can still be bought off with cigarettes!

From this vantage, then, we can see that the deepest logic at work in the novel is a cunning one indeed: each side of the fundamental dichotomy of the discourse of species bifurcates in the novel, *but not along species lines*. For in the end it is the cannibalistic Kigani—not Amy and certainly not the ERTS party—who are most clearly paired with the gray gorillas. Like chimpanzees, they throw feces at their adversaries (308); like the gray gorillas, they will wait until night to attack en masse, only to improbably break off their assault at the moment of victory because of the ERTS expedition's control of the directionality of mimesis: in this case, the killing of the Kigani's *angawa* sorcerer who leads the attack. (Kill him, Munro tells us, and the Kigani will—in response to the perceived fateful message from the gods which the action successfully mimes—break off their attack, no matter their prospect of success [308].) Just as Amy, the First World ape, is able to mimetically master the Third World gray gorillas, so it is with the ERTS party's mimetic power over the Kigani. And thus mimetic doubling, and its potential reversibility in the novel, is thoroughly recontained within, and in fact reproduces, the discursive site of colonialism itself. It becomes, as it were, an affair for "humans"—that is, for colonizers—only.

If this last characterization seems a catachresis with regard to Amy, it is one that is entirely to the point, for what we find is that in the end the novel achieves its ideological work by a double articulation. The first-order problem of mimetic confusion is initially displaced onto the category of the animal only (as a site of interchange between Amy and her wild counterparts); and then that confusion is "remedied" by a second-order sorting according to "the organization of mimesis" by the neocolonial project. In this way we can readily imagine a semiotic square in which "human" means "colonizing mimetic primate" and "animal" means "colonized mimetic primate." On one side we find the "humanized humans" of the ERTS party and the "humanized animal," Amy; on the other side are the "animalized humans," the Kigani, and the "animalized animals," the gray gorillas. What is so revealing about this strategy, of course (particularly in a novelist so known for his technophilia, use of scientific fact, and fascination with big science), is how "unscientific" the whole procedure is—one is tempted to say, in the old Marxist sense, how "ideological." But as we have seen (and as Horkheimer and Adorno themselves realized) Marxist tools are not enough here,[19] for such a procedure also shows us the value of a Deleuzean-Guattarian skepticism toward issuing ethical warrants based on identity achieved rather than multiplicity respected, an identity that in Crichton's novel is purchased at a political price levied in another part of the field, as Amy's mobility and transcendence are paid for by the "animalizing" of the Kigani.

I want to end by suggesting that if *Congo* rearticulates the discourse of species on the more fundamental discursive site of neocolonialism, this does not mean, here as in Hemingway's *Garden of Eden*, that theoretical and ethical problems raised by the discourse of species are ever simply *reducible* to the problematics of race or nation. That has been the point, after all, of my insistence throughout these pages on the *institution*, not simply the discourse, of speciesism. In this light, it is not at all clear that the sort of postcolonial critique we find in Bhabha's work, for example, is of much help in the larger critical project of confronting the question of species difference—despite its immense value in helping us complexify our understanding of how the discourses of species and colonialism interact in sometimes unexpected ways. For Bhabha's work remains captivated, I think, by the figure of the human and suggests that a proper sorting of the subjects in Crichton's *Congo* would proceed rather conventionally along species lines, restoring the properly "ambivalent" dialectical link between the ERTS party and the Kigani (who might then engage in the doubling process of colonial mimesis we have already discussed), while Amy and the gray gorillas would be relegated to the silence they inhabit in the very Enlightenment discourse so forcefully critiqued by Bhabha himself. For a reason we have been revisiting throughout this book—namely, that the discourse of species has been used historically as a chief strategy for marking and exploiting other *human* subjects as well—this strikes me as an important lacuna in any critique of colonialism.

The chief symptom of this residual humanism in Bhabha's work is the crucial role played by what he calls "the performativity of [cultural] translation as the staging of cultural difference" (227), a process of negotiation between the discourse of the colonizer and the colonized in which "the differential systems of social and cultural signification" produce "the foreign element in the midst of the performance of cultural translation" (227). For Bhabha, it is this process of "cultural translation" that "opens up an interruptive time-lag in the 'progressive' myth of modernity, and enables the diasporic and the postcolonial to be represented. But this makes it all the more crucial to specify the discursive and historical temporality that interrupts the enunciative 'present' in which the self-inventions of modernity take place" (240). The aim, Bhabha writes, is "to establish a *sign of the present*, of modernity, that is not that 'now' of transparent immediacy" familiar to us from the Enlightenment myth of progress and its synchronous vision of historical time. This "time-lag," as Bhabha puts it, "is not a circulation of nullity, the endless slippage of the signifier or the theoretical anarchy of aporia"; rather, "it is the problem of the not-one, the minus in the origin and repetition of cultural signs in a doubling that will not be sublated into a similitude. What is *in* modernity *more* than modernity is this signifying 'cut' or temporal break" (245). "This transvaluation of the symbolic structure of the cultural sign," Bhabha concludes, "is absolutely necessary so that in the renaming of modernity there may ensue that process of the active agency of translation—the moment of 'making a name for oneself'" (242).

The question raised by the indispensability of cultural translation in Bhabha's work is whether such a model can do justice to the ethical and theoretical challenges raised by the question of nonhuman animals. More pointedly, the issue is whether the by-definition silence of nonhuman others under Bhabha's model of "cultural translation" is equated with an absence of subjectivity *tout court*, whether such a model can even enable the question to arise. And if this characterization is apt, then how do we take account of, say, the ethical and theoretical challenge we face when we come face to face with Sue Savage-Rumbaugh's Kanzi, for whom "making a name for oneself" in Bhabha's sense seems to be utterly beside the patronymic point (or so Deleuze and Guattari would insist)? For once we have rewritten the figure of the human in Bhabha's terms, it is *still* necessary to understand that if the colonized opens up a "time-lag" in relation to the colonizer's modernity, then the nonhuman other, we might say, *is even slower than that* in relation to both those forms of the human, insofar as a radically different form of experience and temporality is introduced by it—and with a "foreignness" that makes Bhabha's colonial negotiations look rather like an in-house affair by comparison.

To put it another way, it is not clear how it would be wrong to say of Bhabha's work that it fuses, under the figure of the "human," the right not to be colonized with the ability to engage in "cultural translation." And insofar as this assessment is accurate, it seems that Bhabha's work reinstates the discourse of speciesism (if only by implication) to reproduce an image of the colonized as imagined by the colonizer—as one who is mute, whose mimetic ability produces not excess, ambivalence, and reversibility but rather fidelity. Only here, the colonized is not the human other of the Third World—the Kigani, say—but rather any and all nonhuman animals, regardless of their "location." It would be too pointed, perhaps, but it would also not be wrong, to say that Bhabha's work stands in relation to the gray gorillas as Crichton's does to the Kigani. This is to suggest not that Bhabha is wrong, but rather that he is only half right.

Conclusion

Postmodern Ethics, the Question of the Animal, and the Imperatives of Posthumanist Theory

For many readers, I hope, the foregoing pages will constitute a beginning, an opening: an invitation to explore in their own critical practice what it would mean in both intellectual and ethical terms to take seriously the question of the animal—or the *animals*, plural, as Jacques Derrida admonishes us. I hope it is obvious by now that I have had no intention in this book of providing a "foundation" on which we might justify more humane, less exploitive treatment of nonhuman animals. I say this not because the question is not very pressing; indeed, I think it entirely possible, if not likely, that a hundred years from now we will look back on our current mechanized and systematized practices of factory farming, product testing, and much else that undeniably involves animal exploitation and suffering—uses that we earlier saw Derrida compare to the gas chambers of Auschwitz—with much the same horror and disbelief with which we now regard slavery or the genocide of the Second World War.

To be sure, I have suggested (in my discussion of Luc Ferry's *New Ecological Order*) that we ought to be nondiscriminatory with regard to species (just as we are with regard to race or gender, among other qualities) in recognizing the characteristics and potentialities that are widely agreed to constitute what animal rights philosopher Tom Regan calls "the subject of a life." This does not mean, however (as I argue in the opening chapter), committing ourselves to a form of naturalism in ethics, as if there were some sort of linear or transparent relation between ethical concepts (which are necessarily human and social) and the objects in the world toward which they are directed. As Derrida reminds us—and here he would be joined by every poststructuralist theorist I can think of—there can be no "science" of ethics, no "calculation" of the subject whose ethical conduct is determined in a linear way by scientific discoveries about animals (or anything else). My earlier point about this is an entirely pragmatic one, and it is perhaps best summed up, epistemologically speaking, by Richard Rorty:

> By getting rid of the idea of "different methods appropriate to the natures of different objects" (e.g., one for language-constituted and another for non-language-constituted objects), one switches attention from "the demands of the object" to the demands of the purpose which a particular inquiry is supposed to serve. The effect is to modulate philosophical debate from a methodologico-ontological key into an ethico-political key.[1]

If it is true that "it is contexts all the way down," that "we can only inquire after things under a description, that describing something is a matter of relating to other things, and that 'grasping the thing itself' is not something that precedes contextualization" (100), then, as Rorty succinctly puts it, "*holism takes the curse off naturalism*" (109). My point about this in chapter 1 is that it is precisely the *absence* of such a representationalist relation that makes it so important to maintain a procedural rigor and consistency in recognizing those attributes, *wherever we find them*. Precisely *because* the rules of the game are ungrounded, unchecked, and uninsured by anything else, it is all the more critical that we not abandon them when the judgments they require bump up against our most deep-seated prejudices—in this case, prejudices based on species.

From this perspective—the pragmatist perspective that insists that the immanence of such procedures always be immanent *enough*—undertakings such as the well-known Great Ape Project (spearheaded by Peter Singer and Paola Cavalieri and supported by such scientists as Jane Goodall, Roger and Deborah Fouts, Richard Dawkins, Jared Diamond, and the late Carl Sagan, among others) is not only understandable but in fact long overdue. This declaration of basic, universal rights (the "Right to Life, Protection of Individual Liberty, and Prohibition of Torture") not only for human beings but also for chimpanzees, gorillas, and orangutans,

> urges that in drawing the boundary of this sphere of moral equality, we should focus not on the fact that we are human beings, but rather on the fact that we are intelligent beings with a rich and varied social and emotional life. These are qualities that we share not only with our fellow humans, but also with our fellow great apes. Therefore, we should make membership of this larger group sufficient entitlement for inclusion within the sphere of moral equality. We seek an extension of equality that will embrace not only our own species, but also the

species that are our closest relatives and that most resemble us in their capacities and their ways of living.[2]

And yet here, as I have already suggested in my discussion of Singer's animal rights philosophy, one can easily glimpse some of the problems involved in approaching the question of ethics and animals in purely pragmatic or immanent terms. I don't want to rehearse that entire discussion; I will merely note that the model of rights being invoked here for extension to those who are (symptomatically) "most like us" only ends up reinforcing the very humanism that seems to be the problem in the first place. To put it very telegraphically, great apes possess the capacities that we possess, but in diminished form, so we end up ethically recognizing them not because of their wonder and uniqueness, not because of their difference, but because they are inferior versions of ourselves, in which case the ethical humanism that was the problem from the outset simply gets reinforced and reproduced on another level. Now it's not humans versus great apes, its humans and great apes—the "like us" crowd—versus everyone else.

In these terms, then—to put it as bluntly as I know how—I am happy, practically speaking, to support the Great Ape Project (or the revision and upgrading of the United States Animal Welfare Act, or any number of other similar initiatives); and this "practically speaking" should not be taken as a gesture of derision and demotion, since the stakes here—as everyone from Derrida at one end to Singer at the other has noted—are frighteningly high for millions, even *billions*, of nonhuman animals even as I speak. But I offer such support only in abeyance, as it were, only in recognition of the underlying fact that the operative theories and procedures we now have for articulating the social and legal relation between ethics and action are inadequate—and here is the full posthumanist force of the question of the animal in this connection—inadequate for thinking about the ethics of *the question of the human as well as the nonhuman animal.* Practically speaking, we must use what we have, in the same way that one might very well want to invoke the discourse of universal human rights to prohibit the torture of human beings, even though in theoretical terms the model of universal human rights has been thoroughly dismantled as a very historically specific relic of Enlightenment modernity—and for many of the same reasons as in the foregoing critique of *animal* rights.

What this means, then, is that such projects, which strategically invoke ethical models and theories that are rhetorically very powerful precisely *because* they are relics, because they are "residual" (to use Raymond Williams's well-worn term), are in fact, in intellectual terms, the *easy* part. What is harder, I think, is the project I have tried to make a start on in this book: to address, squarely within the purview of postmodern theory, the theoretical and ethical complexities that attend the question of the animal in several registers. This means considering not only that we share our world with nonhuman others who inhabited this planet before we arrived on the scene and will in all likelihood far outlast the tenure of *Homo sapiens* but also that *we*—whoever "we" are—are in a profound sense constituted as human subjects within and atop a nonhuman otherness that postmodern theory has worked hard to release from the bad-faith repressions and disavowals of humanism—whether in Deleuze and Guattari's invocation of the multiplicity of the subject "becoming-animal" in their critique of psychoanalysis, in Derrida's insistence on the fundamentally "inhuman" quality of language itself and the subjection of "the living" in general to the force of the trace, in Donna Haraway's focus on the multiplicity and situatedness of the subject, and in myriad other ways.

As even this brief sampling suggests—and I touched on this in the introduction—the theoretical and ethical issues that attend the question of the animal are only part of the larger issue of nonhuman modes of being and are therefore inseparable (for this reason and for others that I will take up in a moment) from the broader challenge of posthumanist theory—a challenge whose theoretical lineaments I attempted to articulate in some detail in my previous book, *Critical Environments: Postmodern Theory and the Pragmatics of the "Outside."*[3] There I investigated how prominent varieties of contemporary theory—chiefly pragmatism, poststructuralism, and systems theory—confront the problem of thinking about the "outside," the "not-Me" (as Emerson once put it), the domain of what used to be called Nature and the *object.* And in this light the current book constitutes for me not only a beginning but also an end, a completion, the second half of a project begun in that earlier book. It similarly undertakes an investigation of the conditions of possibility for posthumanist theory—and what readings of literature, film, and culture underwritten by that theory might look like—but it does so this time on the terrain of the "inside," the site of what used to be called the "self" and the "subject." In both halves of the project, my premise has been that maintaining a commitment to distinctly posthumanist ways of theorizing the questions at hand (no matter how counterintuitive or implicated in what is sometimes defensively called theoretical "jargon") will enhance our understanding of the embeddedness and entanglement of the "human" in all that it is not, in all that used to be thought of as its opposites or its others. For now, at the beginning of the twenty-first century, these aspects suddenly seem—sometimes exhilaratingly, sometimes terrifyingly—"in the subject more than the subject itself," to borrow Slavoj Žižek's phrase.

We can clarify our sense of what a commitment to posthumanist theory means for ethics—and how even the most ambitious attempts to theorize what I have called a distinctly postmodern ethical pluralism continue to be plagued by an updated but fundamental humanism—by turning to a brief examination of what is surely one of the most magisterial studies in this area, Zygmunt Bauman's *Postmodern Ethics*. Bauman begins with a distinction that can help explain how we can invoke, as I did a moment ago, a pragmatic domain of ethics, only to turn around and insist that this "actually existing" realm of ethics is in some fundamental sense deeply flawed and inadequate to the questions it seeks to address. As Bauman notes (and this will already be familiar to us from Derrida's distinction between "Law" and "law," justice and law), we must begin by separating the question of *ethics* from the question of *morality*, disengaging political and legal codes of conduct in all their social and historical contingency from the broader philosophical question of the good and the just.[4] Here Bauman rightly argues that postmodernism, which is often associated with "the celebration of the 'demise of the ethical,' of the substitution of aesthetics for ethics and of the 'ultimate emancipation' that follows" (2), is in fact a positive development in allowing us to reopen the questions of the moral and the just that were foreclosed by Enlightenment modernity. "The novelty of the postmodern approach to ethics," he writes, "consists first and foremost not in the abandoning of characteristically modern moral concerns, but in the rejection of the typically modern ways of going about its moral problems (that is, responding to moral challenges with coercive normative regulation in political practice, and the philosophical search for absolutes, universals and foundations in theory)" (3–4). "The moral thought and practice of modernity was animated," he continues, "by the belief in the possibility of a *non-ambivalent, non-aporetic ethical code*," whereas "it is the *disbelief* in such a possibility that is *post*modern" (9–10).

For Bauman, then, "morality is incurably *aporetic*"; "virtually every moral impulse, if acted upon in full, leads to immoral consequences" (as when "care for the Other, when taken to its extreme, leads to the annihilation of the Other, to domination and oppression") (11). And from this several consequences follow. First, "moral phenomena are inherently 'non-rational'"; they therefore cannot be "represented as *rule-guided*" or "exhausted by any 'ethical code'" (11). Second, morality is therefore "*not universalizable*" (12). This is not meant to endorse "the popular opinion and hot-headed 'everything goes' triumphalism of certain postmodernist writers" (14) (as Bauman hotheadedly puts it!), because, third, "moral responsibility—being *for* the Other before one can be *with* the Other—is the first reality of the self, a starting point rather than a product of society." It is, as Bauman puts it altogether unabashedly, "the ultimate, non-determined presence, indeed, an act of creation *ex nihilo*, if ever there was one" (13). For readers who have followed the discussion of Derrida and Lévinas in chapter 2 of this book, the deconstructive cast of these formulations will be familiar; and indeed, Bauman himself announces that the "moral unity is thinkable, if at all, not as the end-product of globalizing the domain of political powers with ethical pretensions, but as the utopian horizon of deconstructing the 'without us the deluge' claims of nation-states" (14).

Here, however, we would do well to bring this "deconstructive" position of Bauman's into sharper focus to understand its relevance and promise for thinking about ethics in relation to the question of the animal. On this point *Postmodern Ethics* seems at first glance particularly promising, especially in its rejection of ethical models that rely on "reciprocity" and "contractualism." What unites them, Bauman writes—and here we will find a very strong reprise of Derrida's discussion in texts such as "'Eating Well'"—is that they "imply *calculability* of action" (59). "What more than anything else sets the contractually defined behaviour apart from a moral one," he lucidly observes, "is the fact that the 'duty to fulfil the duty' is for each side dependent on the other side's record. . . . It is, so to speak, in the power of my partner to set me (by design or by default) 'free,' to 'unbind' me from my duties. My duties are *heteronomic*." It is true, he continues, that "the entering of a contract may be depicted as the expression of my status as an independent decision-maker. From then on, however, there is 'nothing personal' about my actions. Once bound by contract, my actions are 'remotely controlled' by punitive sanctions, administered by the agencies of enforcement" (59).

This is a rather different critique of ethical contractualism than we found in chapter 1 in the work of animal rights philosopher Tom Regan, but it has the similar effect of reopening the ethical relationship to the possibility of nonhuman others. And that promise seems only to be undergirded by Bauman's even more foundational (if one may use the term) critique of the related idea that "reciprocity" is central to the ethical (or in his terms, "moral") relationship. Reciprocity ignores the fact that "'we' becomes a plural of 'I' only at the cost of glossing over the I's multidimensionality" (48) (or what I have elsewhere in this book called the "verticality" of the subject's difference that might be thought of along the lines of Deleuze and Guattari's "becoming-animal," Derrida's "the animal that therefore I am," or Haraway's "situated" subject). "My relation to the Other is *programmatically* non-symmetrical," Bauman continues; "that is, not dependent on the Other's past, present, anticipated or hoped-for *reciprocation*."[5] It is "an essentially *unequal* relationship" (48), in which the other is conceptualized along the lines familiar to us from debates in environmental ethics about our debt to future generations: as "weak, vulnerable, without power; they are indeed without power since they cannot repay what has been done to them (nor for that matter reward our deeds), and vulnerable since they cannot prevent us from doing whatever we think worth doing" (219–20).

It appears that such formulations, in contrast to the dominant ethical models of contractarianism and reciprocity, would be extremely promising for rethinking the question of ethics (or "morality," in Bauman's terms) in relation to nonhuman animals. Unfortunately, that promise is foreclosed by a countervaling humanism in Bauman's work that fatefully determines who can be the subject of moral address, even as it seems that in principle any such limitation would by definition violate the very moral posture that welcomes the other in its absolute otherness. To put it another way, the promising open-endedness of a certain kind of ethical *relation* in Bauman is foreclosed by a certain kind of ethical *quality*—what Bauman simply calls the "moral conscience"—that is the sole domain of a certain kind of *subject*. On the one hand, Bauman writes,

> the Others of the moral relationship are the others we live *for*. These others are resistant to all typification. As residents of moral space, they remain forever specific and irreplaceable; they are not specimens of categories, and most certainly do not enter the moral space in virtue of being members of a category which *entitles* them to be objects of moral concern. They become objects of a moral stance solely by virtue of having been targeted directly, as those concrete others out there, by moral concern. (165)

So far so good; and in fact one might well assume that such a passage could have been lifted from Derrida's "The Animal That Therefore I Am," where he warns, for example, against the typification of nonhuman animals in his protest, "*the* animal, what a word!"

At the same time, however, we find in Bauman the very sort of typification and generic classification that his ethics seem to guard against at all costs—in this case, by means of the designation "humanity" as the exclusive membership of the moral community. Take, for example, the following passage, which condenses many of these promises and shortcomings:

Being a moral person means that I *am* my brother's keeper. But this also means that *I* am my brother's keeper whether or not my brother sees his own brotherly duties the same way I do. . . . At least, I can be properly his keeper only if I act *as if* I was the only one *obliged*, or even likely, to act this way. . . . It is this uniqueness (not "generalizability"!), and this non-reversibility of my responsibility, which puts me in the moral relationship. This is what counts, whether or not all brothers of the world would do for their own brothers what I am about to do. (51)

Such a passage makes it clear that the asymmetry and unaccountability of the moral relationship—about which Lévinas, Derrida, and Lyotard would all agree—is undercut by the countervailing prescription of *who can in principle be party* to that relationship. But to readers who have been listening attentively thus far to Bauman's language, this should come as no surprise. Bauman's subject of the moral relationship—and he is unabashed about this—is wholly enframed by Lévinas's "humanism of the other man" and its presumption that the moral party is always already the subject of the "Face," the one in whose eyes we can discern, however dimly, an awareness of his or her own mortality. Only then, as John Llewelyn reminded us in chapter 2, does the "first word" of the ethical call addressed to me by the other—"Thou shalt not kill"—bind me in a moral relationship. But one might well ask, with Derrida, why this should be the case. If the "passivity," "weakness," and what Derrida calls "this non-power at the heart of power" are central to the unaccountability and asymmetry of the ethical relationship—its ex nihilo and unmotivated character—invoked by Bauman, then why should this on principle exclude the animal other, since, as Derrida puts it, "Mortality resides there, as the most radical means of thinking the finitude that we share with animals, the mortality that belongs to the very finitude of life, to the experience of compassion, to the possibility of sharing the possibility of this non-power, the possibility of this impossibility." The answer, as we saw in Llewelyn's analysis of Lévinas's debt to Kant, is that it is not this finitude and passivity but rather the reflective relation to it via reason and the ability to "universalize his maxim" that is central to the subject of ethics.[6]

What we find, then, in both Bauman and Lévinas, is a domestication and humanization of the more general problems of finitude and alterity in relation to ethics. One the one hand, as we saw in Richard Beardsworth's discussion in chapter 2, both Lévinas and Derrida take issue with Heidegger because he "appropriates the limit [of death] rather than returning it to *the other* of time. The existential of 'being-towards-death' is consequently a 'being-able' (*pouvoir-être*), not the impossibility of all power"[7]—not, that is, the "passivity" that both Lévinas and Derrida place at the center of the ethical relationship. And Lévinas shares with Derrida a commitment to our fundamental submission to the essential *inhumanity* of time, in which death is neither "for" me *nor* for the other (since it absolutely exceeds the experience of each). It is the very form of the alterity of time itself that, as Beardsworth puts it, "rather than signalling the other signals the *alterity* of the other" (132), the "there" from which the impossible call of the other comes for Lévinas, the very (non)space of what Derrida calls the impossible "promise" that is always deferred, always "to come." But in giving precedence to the human form of alterity over others (the other in the singular over others, man over what Derrida calls "the living," the Jewish man over humanity in general, and so on), Lévinas, as Beardsworth puts it, loses "the aporia of law by surrendering a differentiated articulation *between* the other and the same, between ethical relation and temporalization." And "the effect of this loss is the loss in turn of the *incalculable* nature of the relation between the other and its others" (125).

It is this incalculability—this multiplicity, if you like—that is meant to be marked and kept permanently open by Derrida's insistence not just on the question of the animal, but on the *animals*, in the plural. To put it in terms that combine both the ontological and formal registers, Lévinas reestablishes *a human rather than an inhuman relation to the inhumanity of time* (and at the same stroke cordons off the animal other from the passivity it indicates) by fixing the other (little *o*, alterity in its incalculable *différance*, its "iteration," in Derrida's terms) as the Other (big *O*) "as such"—a fixation that gets thematized and reontologized as the Other-as-Human. In which case, Beardsworth argues, the "there" of time's essential inhumanity, in which resides the all-important passivity that makes ethics possible, "is thereby disavowed and the other humanized. It is ultimately in this humanization of the other of time, consequent upon the disavowal of the radical inhumanity of the there, that the other is justified" (141).

And if this is the case, then the "unaccountability" or "incalculability" that is absolutely central to the moral relationship—the "without reason" that is a "strategy of opening, which breaks the monadic immanence and makes the subject into one-that-steps-outside-of-itself, the subject of self-transcendence"[8]—is not so unaccountable after all. To put it another way, the humanism (or "modernism," to use Bauman's preferred term) that plagues the generalizability and universalism of ethical codes and their prescriptives, which Bauman rejects in favor of nongeneralizable moral responsibility, is here displaced onto *who can be* the addressee of the ethical "call" in the first place. Even as reciprocity is correctly declared beside the point of moral responsibility, Bauman's postmodern ethics takes away with one hand what it gives with the other by making it clear that the *ability to reciprocate* is in fact crucial to membership in the moral community—and this is directly in line with Lévinas's own Kantian position. In other words, Lévinasian ethics may not be based on reciprocity—quite the contrary—but the rejection of reciprocity as the fundamental first move of the ethical relationship gets its force from the assumption of a subject who *can* reciprocate to begin with. But—and this is clear in Bauman's discussion of the structure of the gift (57 ff.), which could well be expanded along Derridean lines—if reciprocation is itself constitutively impossible and indeed in some fundamental sense undermines the true "selflessness" of the moral impulse, then why should this carry any force? Why should not the supremely moral act be that directed toward one, such as the animal other, from whom *there is no hope, ever, of reciprocity?*[9]

From another, more strictly epistemological vantage, a fundamental problem with Bauman's formulation of the moral relationship is its essentially circular, tautological nature—or I should say, more precisely, its failure to make that circularity and tautology productive, a failure that therefore thrusts Bauman's avowed postmodernism back into the very modernist humanism he abjures. Here Lévinasian humanism joins with humanism of a rather more traditional sort, as we find in *Postmodern Ethics* many pronouncements of the sort that moral responsibility is "'from the start,' somehow rooted in the very way we humans are" (34), that "it must be the moral capacity of human beings that makes them so conspicuously capable to form societies and against all odds secure their—happy or less happy—survival" (32), or—and this is how the book ends—that "moral responsibility does not look for reassurance for its right to be or for excuses for its right not to be. It is there before any reassurance or proof and after any excuse or absolution" (250). Most startling of all, perhaps, particularly in conjunction with the foregoing quotation, is Bauman's assertion that "the moral crisis of the postmodern habitat requires first and foremost that politics—whether the politics of the politicians or the policentric, scattered politics which matters all the more for being so elusive and beyond control—be an extension and institutionalizaton of moral responsibility" (246).

But why not, after all, say exactly the reverse? Why not say, along with Bauman's fellow sociologist Niklas Luhmann, that it is precisely the moralization of politics—*particularly* if the moral is situated beyond "reassurance or proof," above "any excuse or absolution"—that is the danger? Wouldn't the perils of such a moralization of politics be especially palpable for those who lived through the Reagan era and the Moral Majority during the 1980s or, more recently, the transcoding of patriotic fervor in terms of Christian religious doctrine in the wake of the 9/11/01 attacks on New York and Washington, as billboards everywhere flatly declared the homology: "In God We Trust. United We Stand"? To articulate the differences and what is at stake in them, it is important to understand that Luhmann begins with the same sort of description of postmodern society that we find in Bauman but draws from it very different conclusions. For Luhmann, as I have discussed in some detail elsewhere, what above all characterizes the evolution of modern society (and its intensification under postmodernity) is "functional differentiation," the distribution of society's operations across a horizontally distributed, nonhierarchical set of subsystems (the educational system, the economic system, the legal system, and so on), each operating and reproducing itself by means of its own code, with none being dominant.[10] Like the "language games" of Wittgenstein and Lyotard, these codes are fundamentally incommensurable, and there is no Archimedean point from which they can all be viewed at once, as "homologies" of each other in some totality. In fact, it is this incommensurability that guarantees that the subsystems can carry out their functions so that, for example, a legal decision is based not on the code of knowledge/ignorance (education) or profitable/not profitable (economic)—in which case those committing intelligent or profitable crimes could never be found guilty!—but rather on the code legal/illegal. This functional differentiation allows the subsystems to better handle the increasingly complex environment in which they find themselves by means of a kind of dampening or filtering made available by the reduction of overwhelming complexity via systemic coding—but only at the expense, of course, of increasing the complexity of the environment for the *other* subsystems. As the legal system builds up its own internal complexity to handle changes in its environment (think, for example, of the fitful extension of copyright law in relation to digital forms of reproduction such as Napster), it in turn increases the complexity of the environment in which the economic system, for example, must operate and adapt.

As part of this evolution, the moral code for Luhmann, as William Rasch puts it, "has detached itself from its premodern locus in religion and has become a self-replicating, parasitic invader of the various modern, functionally differentiated social systems," with the attendant danger that the moral code is now more or less free-floating and may attach itself "isomorphically" to the codes of the various function systems "to impose a binding translation of 'true' or 'government' or 'profitable' into 'good' (or 'bad')."[11] The fact that this *is* a danger—and it is one that has particular resonance for a German intellectual like Luhmann working in the post–World War II period[12]—"can be elucidated historically," Rasch writes, "by listing the countless crusades, wars, inquisitions and persecutions that moral discourse has fueled" (93). Once morality becomes "unhoused" from its location in the church with the emergence of functionally differentiated society, it is the function of *ethics* to serve as "a kind of *immune system* or *on/off switch*" that "emerges as the by-product of a system's attempt to preserve its own reproduction from the ravagings of moral infection" (94). "Ethics becomes formalized," as Rasch puts it, and questions of good and bad are "moved from a consideration of the moral 'fiber' or substance of an individual to a consideration of action in the face of competing alternatives" (93). In this way ethics attempts to ensure the autopoiesis and autonomy of the various function systems by limiting and, if one likes, relativizing morality.

Rasch argues that this insistence on the insulation of the various function systems and their autopoiesis from the totalizing effects of the moral code makes possible a more nuanced understanding of contemporary political phenomena, such as Bill Clinton's simultaneously opposing abortion on moral grounds and affirming a woman's legal right to choose in his 1992 Democratic National Convention acceptance speech. If we judge such a statement in terms of the moral code's parasitical totalization and think only in terms of the "moral fiber," "authenticity," "sincerity," and so on of the subject who utters such words, then such a statement will likely strike us as hypocritical and dishonest. But from the vantage of an understanding of the political system in the broader context of functional differentiation, "it could also be construed as distinctly 'modern' in the sense of a radical—and radically desired—disjuncture between legal and moral codes" (92).[13] Such a disjuncture would help to explain the central fact about the Clinton presidency that drove the political Right crazy throughout the 1990s: that even as a vast majority of the American public condemned Clinton's moral conduct in the wake of the Monica Lewinsky affair, he at the same time enjoyed historically unprecedented approval ratings by the American public for his execution of his duties as president. And it would allow us, moreover, to rearticulate the distinction I invoked at the outset: between the pragmatics of using existing legal codes ("rights") to respond to changes in the legal system's larger environment with regard to what we know about animals, their phenomenology, and so on (produced in this case chiefly, but not solely, by another subsystem: science) while at the same time arguing from within a different code (ethics) that such models are inadequate.

Of course Bauman would no doubt see all of this as exemplary of the submission to the mere "roles" made available to us by society's various ethical codes that only evade the question of morality itself. In many places in

Postmodern Ethics, Bauman invokes this sort of distinction. "Our life work," he writes, "is split into many little tasks, each performed in a different place. . . . In each setting we merely appear in a 'role,' one of many roles we play. None of the roles seems to take hold of our 'whole selves,' "'what we truly are,'" "our 'real selves'" (19). "There are too many rules for comfort," he continues; "they speak in different voices. . . . They clash and contradict each other, each claiming the authority the others deny. It transpires sooner or later that following the rules, however scrupulously, does not save us from responsibility. After all, it is each one of us on his or her own who has to de-cide" (20). Of course, what Bauman here decries as morally tortuous, Luh-mann has already anatomized as simply unavoidable in functionally differ-entiated society.

But the more fundamental issue here, as I have already suggested, is that such formulations—and they are hard-wired to the entire project of *Post-modern Ethics*—are utterly question-begging, as the scare quotes around "real selves," "whole selves," and the like readily communicate. They as-sume, of course, that there *is* some untainted, unmediated space of subjec-tive interiority where one can stand aside from, totalize, and critique these codes and roles without at the same time being bound by them, some un-questioned and unquestionable yardstick by which the socially and linguis-tically constructed codes of ethics may be judged—otherwise the entire force of the distinction between morality and merely ethical "roles" is lost. It goes without saying, I think, that such a move—what Richard Rorty has humorously called the "God's-eye standpoint" (6)—has been thoroughly discredited by contemporary theory since the 1960s, and here one could cite any number of sources: Lacan's rereading of Freud in light of the realization that "the unconscious is structured like a language," that it is *outside* the sub-ject; Derrida's deconstruction of the "auto-affection" and self-presence of voice in both Husserl and Saussure; Althusser's dismantling of humanist Marxism; or beyond that, Lévi-Strauss's critique of Sartre's neo-Hegelian reliance on the category of consciousness. As I have argued in some detail in *Critical Environments,* if we have learned anything from theory over the past three decades, it is that there is no "before" of the symbolic, of language, and the social, whether this is construed in terms of the domain of the subject (philosophical idealism's gambit) or of the object (realism and positivism)— which is not to say, by a long shot, that language and the symbolic are "all there is," as Richard Rorty has perhaps most cogently demonstrated.[14]

The presumption of such a space of pure interiority, pure self-presence of the subject, receives its most rigorous and devastating critique, perhaps, in Derrida's reading of Husserl in *Speech and Phenomena,* which would no doubt draw our attention to the demotion of writing and iteration thinly veiled (if at all) in Bauman's privileging of "moral conscience" over the realm of the ethical. My aim here, however, is not to retrace any of these arguments, but simply to point out that all we get by way of confrontation with this question in *Postmodern Ethics* are passages like the following: "Morality is 'before being' only in its own, moral sense of 'before': that is, in the sense of being 'better.' But in the ontological sense, the sense which gets the upper hand whenever the two senses compete in the realm of being, the realm we are all in—being is before morality" (75). Or again, "The language we use (the only one we can use) is a sedimentation of life organized under the auspices of ontology's unchallenged domination. It is a language shaped to report and to account for *being*, construed the way ontology defines it; the concept of 'being' and all its correlates and derivatives convey matter-of-factly ontology's right to define. One may be helped to struggle through and away from the resulting difficulty by remembering that in Lévinasian ethical discourse 'being' appears, as Derrida would say, *sous rature*" (72–73 n. 12). My point here, of course, is that while Bauman wants to deploy the concept of *sous rature* to bolster the distinction between the moral and the ethical, he relies on the very kinds of assumptions about the self that Derrida's work—on erasure, on the trace, on Husserl, on writing versus speech—is meant to dismantle.

This failure is crucial, not only because Bauman's distinction between the moral and the ethical thereby founders, but also because his attempt to confront the problem—or fashion, if one likes—of postmodern "relativism" thereby runs aground as well—a problem that it is critical to confront head-on, as I argue in *Critical Environments*, if we are to work through the difficulties of theorizing a distinctly postmodern ethical and political pluralism.[15] In the meantime, Bauman is caught in the same paradox in Lyotard's work that I discussed in chapter 2, one pointed out by Samuel Weber in his afterword to *Just Gaming*: namely, that all claims and discourses are here subject to the "deconstructive" and "open-ended" force of the moral—except the claim for the priority of the moral itself, the questioning of which would be, paradoxically, both deconstructive *and* immoral. So while Bauman may have a desire for a concept of the "moral" that evades such problems by being aporetic and deconstructive, he does not have a *theory* for it. And in the absence of such a theory, proclamations of the need to make politics a function of morality can only appear incoherent or worse.

We can get a clearer sense of the epistemological problems here and how they might be thought through more productively, I think, by once again glancing at Luhmann's work. Luhmann insists that all observations—which are all, formally speaking, communications—are based on system codes, which are themselves built on a constitutive distinction that is essentially paradoxical or tautological. Moreover, an observation cannot acknowledge that distinction's tautological nature *and* at the same time mobilize that distinction to carry out its operations. The legal system, for example, cannot reproduce itself and at the same time acknowledge the paradoxical identity of the two sides of the fundamental distinction of the legal code (i.e., the difference between legal/illegal is self-instantiated, comes from nowhere, and takes place on one side of the distinction only; the fundamental tautology that underlies the code's distinction is "legal is legal"). What this means, as Luhmann puts it, is that all such observations and systems are built on an inescapable "blind spot" that only *other* observations, from within *other* systems, can reveal. For Luhmann, however, this blindness does not separate or alienate us from the world but instead ensures our connection with it. In fact, it is from this fundamental and inescapable "blindness" that Luhmann derives the necessity of the other and the world, the "outside" of the system. As he puts it in a rather remarkable passage in the essay "The Cognitive Program of Constructivism and a Reality That Remains Unknown,"

> The source of a distinction's guaranteeing reality lies in its own operative unity [the unity of legal/illegal, for example]. It is, however, precisely as this unity [the paradoxical identity of its difference] that the distinction cannot be observed—except by means of another distinction which then assumes the function of a guarantor of reality. Another way of expressing this is to say the operation emerges simultaneously with the world which as a result remains cognitively unapproachable to the operation.
>
> The conclusion to be drawn from this is that the connection with the reality of the external world is established by the blind spot of the cognitive operation. *Reality is what one does not perceive when one perceives it.*[16]

And what this means, in turn—and here I think we find an articulation very much in line with the Derridean "iteration" of the alterity of the other—is that the inexhaustability of the "outside" world for Luhmann does not reside in some preexisting ontological positivity, substance, or fullness—not in any nature "as such"—but rather emerges *from* the inside, from an observation that is able to cognize, communicate, and make meaning *at all* only by "reentering" the distinction between x and y, inside and outside, on one side of the distinction itself—namely, the inside. As Luhmann puts it, "In a somewhat different, Wittgensteinian formulation, one could say that a system can see only what it can see. It cannot see what it cannot. Moreover, it cannot see that it cannot see this."[17] In rather more familiar terms: The distinction between language and nonlanguage is itself made within language; the cognition of figure and ground requires a frame, the separation of signal and noise, a code. Moreover, because the outside is the outside *of* the inside, the differential or "incalculable" structure of alterity is preserved, in two ways. The inexhaustability of the outside—the fact that it is always "to come," in Derrida's phrasing—resides not only in any first-order observation's contingent (and therefore possibly *different*) act of selection and exclusion, but also in its constitutive incompleteness and paradoxicality, which in turn generate the necessity for *other* second-order observations, using *other* distinctions, which are themselves constitutively paradoxical, and so on. Luhmann calls this process not the "iteration" of the difference between law and Law as we find it in Derrida, but rather the "unfolding" of the paradoxical identity of difference of any given first-order observation in a second-order plurality of horizontally distributed systems that, even though they share a radical equivalence in formal terms, are discrete and discontinuous, since they are neither "moments" in some unfolding dialectical process nor "homologies" of each other in some fully observable social totality. Observation will thus, Luhmann writes,

> maintain the world as severed by distinctions, frames, and forms *and maintained by its severance*. . . . This partiality precludes any possibility of representation or mimesis and any "holistic" theory. . . .
>
> The operation of observing, therefore, includes the exclusion of the unobservable, including, moreover, the unobservable par excellence, observation itself, the observer-in-operation.[18]

To put it schematically, the "exclusion" described here is Luhmann's version of what Derrida identifies as the "sacrificial structure" of all symbolic economies. For both—and this, I think, despite Luhmann's disclaimers[19]—the iteration of this exclusion always produces a necessary outside and other that has a fundamentally ethical force; it is a pluralism that calls into question any given reason, but only by the pursuit of reason itself. And this means—to invoke once again the association of reason with the human and nonreason with the animal in the Kantian legacy that we have seen in both Lévinas and Bauman—that the human makes way for the animal, but only by means of the human itself. Here (and precisely by *increased* pressure on the sorts of paradoxes Bauman simply abandons) we find a theory that can begin to make good on Bauman's understandable suspicion of the sort of pure proceduralism in ethics that one would find in a Rorty or a Stanley

Fish, where ethics simply means, to use Fish's phrase, "doing what comes naturally" in a pure immanence of ethnocentric interpretive communities whose theorizing generates not the necessity of other observations and a subjection to the community's "outside" but precisely their foreclosure.[20] It allows us to respect and sustain Bauman's essentially Lévinasian commitment to keeping open the difference between law and Law—the question of justice beyond any immanence or any existing procedural code—in his insistence that "morality is endemically and irredeemably *non-rational*—in the sense of not being calculable" (60). But it enables us to do so precisely *by means of reason and its iteration*, and not by hypostatizing and reontologizing some mysterious and well-nigh theological "real self" and the "there" of the moral domain.

We might say, then, that Luhmann is to Bauman on the site of sociology what Derrida is to Lévinas on the terrain of philosophy. For both Luhmann and Derrida, the "there" of the outside emerges only as the outside *of* the inside, only by means of the differential iteration of reason-as-*écriture*—or, in Luhmann's terms, the difference between first-order and second-order observation—that, in its irreducible difference, keeps open the alterity of the other and makes possible the ongoing question of the difference between law and Law, observation and what it excludes. "I am moral *before* I think," Bauman writes (61). But the point to be taken from both Luhmann and Derrida is that this "before" always comes *too late*, since there can be no unmediated encounter with or communication of this "before" prior to language, writing, and the social. Hence the status of ethics, justice, and the promise in Derrida as always "to come," as that which exceeds the very iteration through which it is brought into being.

To put it another way, we might say that if Bauman's contention "I am moral *before* I think" may also be rewritten as "I am *human* before I think"— a coupling that his Lévinasian position not only makes possible but mandates—then we could say that Luhmann's point, and Derrida's, is that since this "before" is always the "before" of the "after" of writing and the social, we must contend instead that "I *can* be *inhuman* only *after* I think." Far from hindering a postmodern ethics, however, this is precisely what enables it, since reason and the human, in their attempt to be true to themselves, to "do what they do best," thereby systematically produce the "unaccountable" and "incalculable" excess, outside, and other by which they are interrogated. Paradoxically, then, the most pointed irony of all is that Bauman only further entrenches the very humanism his "postmodernism" was meant to unsettle, not because he defends reason against all its others (including, of course, the animal) but precisely because he *abandons reason too quickly* rather than rigorously pursuing it until it loops back on itself and turns into its other, pointing us toward the "space," as Derrida would say, where concepts fail, the space they point to but cannot encapsulate. Or to put it in Luhmann's terms,

> When observers (we, at the moment) continue to look for an ultimate reality, a concluding formula, a final identity, they will find the paradox. Such a paradox is not simply a logical contradiction (A is non-A) but a foundational statement: The world is observable *because* it is unobservable. Nothing can be observed (not even the "nothing") without drawing a distinction, but this operation remains indistinguishable. It can be distinguished, but only by another operation. It crosses the boundary between the unmarked and the marked space, a boundary that does not exist before and comes into being (if being is the right word) only by crossing it. Or to say it in Derrida's style, the condition of its possibility is its impossibility.
>
> Obviously this makes no sense. It makes meaning. ("Paradoxy," 46)

And it makes it possible to understand that, paradoxically, a truly postmodern ethical pluralism can take place not by avoiding posthumanist theory but only by means of it—not by rushing toward the other, all the others, in some redemptive embrace, but precisely *by way of* theory, by "doing what we do best." My wager here has been that doing it well enough will reveal that, theoretically and ethically speaking, the only way out is through, the only way to "before" is "to come," and the only way to the "there" in which the animals reside is to find them "here," in us and of us, as part of a plurality for which perhaps even "the animals," in the plural, is far too lame a word.

Introduction

1. For the definitive introduction of the term in the animal rights literature, see Peter Singer, *Animal Liberation: A New Ethics for Our Treatment of Animals* (New York: Avon Books, 1975), 7.

2. Throughout this book, the term "animal" should always be taken to mean the more technically accurate, but stylistically infelicitous, term "nonhuman animals." Similarly, for convenience I will use the pronoun "it" to refer to the animal, but I do so recognizing, as Jacques Derrida points out in chapter 2 below, that speaking of the animal with the neuter pronoun is a barometer of our inability to take seriously the possibility of the animal's nongeneric being.

3. See, for example, "Can Animals Think?" *Time* 141, 12 (March 22, 1993): 54–61; "The Secret World of Dogs" and "Not Just a Pretty Face," *Newsweek*, November 1, 1993, 58–61, 63–67; and "What Animals Say to Each Other," *U.S. News and World Report*, June 5, 1995, 50–56.

4. See, for a useful overview, Paola Cavalieri and Peter Singer, eds., *The Great Ape Project: Equality beyond Humanity* (New York: St. Martin's Press, 1993). Marian Stamp Dawkins's *Through Our Eyes Only? The Search for Animal Consciousness* (New York: W. H. Freeman/Spektrum, 1993) is the most recent work by one of the most important researchers in establishing the scientific foundation for understanding animal suffering, and Donald R. Griffin's *Animal Minds* updates the work of, in effect, the founder of the field of contemporary cognitive ethology (Chicago: University of Chicago Press, 1992). And see, in a more scholarly vein, the essays collected in Marc Bekoff and Dale Jamieson, eds., *Interpretation and Explanation in the Study of Animal Behavior*, vol. 1 (Boulder, Colo.: Westview Press, 1990).

5. Donna J. Haraway, *Simians, Cyborgs, and Women: The Reinvention of Nature* (New York: Routledge, 1991), 151–52.

6. Sigmund Freud, *Civilization and Its Discontents*, trans. and ed. James Strachey (New York: Norton, 1961), 51–52 n. 1.

7. Donna J. Haraway, "Situated Knowledges: The Science Question in Feminism and the Privilege of Partial Perspective," in her *Simians, Cyborgs, and Women*, 188.

8. Thomas Nagel, "What Is It Like to Be a Bat?" *Philosophical Review* 83, 4 (October 1974): 435–50. As Nagel observes, "Bats, although more closely related to us than those other species, nevertheless present a range of activity and sensory apparatus so different from ours that the problem I want to pose is exceptionally vivid"— that bats "perceive the external world primarily by sonar, or echolocation," which, "though clearly a form of perception, is not similar in its operation to any sense that we possess" (438). For instructive and sometimes amusing responses to Nagel's essay, see Kathleen Akins, "Science and Our Inner Lives: Birds of Prey, Bats, and the Common (Featherless) Bi-ped," in Bekoff and Jamieson, *Interpretation and Explanation in the Study of Animal Behavior*; Douglas Hofstadter and Daniel C. Dennett, eds.,

The Mind's I: Fantasies and Reflections on Self and Soul (New York: Basic Books, 1981), 427–38; and Daniel Dennett, *Consciousness Explained* (Boston: Little, Brown, 1991), 441–55.

9. Vicki Hearne, *Adam's Task: Calling Animals by Name* (New York: Random House, 1987), 79. Further references are given in the text.

10. As they put it, "These animals invite us to regress, draw us into a narcissistic contemplation, and they are the only kind of animal psychoanalysis understands, the better to discover a daddy, a mommy, a little brother behind them." See Gilles Deleuze and Félix Guattari, *A Thousand Plateaus: Capitalism and Schizophrenia*, trans. Brian Massumi (Minneapolis: University of Minnesota Press, 1987), 240.

11. Bruno Latour, *We Have Never Been Modern*, trans. Catherine Porter (Cambridge: Harvard University Press, 1993), 136.

12. See Georges Bataille, *Theory of Religion*, trans. Robert Hurley (New York: Zone Books, 1992), esp. 17–61; Jacques Derrida, "'Eating Well,' or The Calculation of the Subject: An Interview with Jacques Derrida," in *Who Comes after the Subject?* ed. Eduardo Cadava, Peter Connor, and Jean-Luc Nancy (New York: Routledge, 1991), 96–119; Jacques Derrida, "Force of Law: The 'Mystical Foundation of Authority,'" *Cardozo Law Review* 11, 919 (1990): 919–71. For a fuller discussion of these specific issues, see chapter 3 below.

13. Gayatri Chakravorty Spivak, "Remembering the Limits: Difference, Identity and Practice," in *Socialism and the Limits of Liberalism*, ed. Peter Osborne (London: Verso, 1991), 229.

14. Toni Morrison, *Playing in the Dark: Whiteness and the Literary Imagination* (New York: Random House, 1992), 44–45.

15. Carol Adams, *The Sexual Politics of Meat: A Feminist-Vegetarian Critical Theory* (New York: Continuum, 1990); Derrida, "'Eating Well,'" 113.

16. See Cary Wolfe, "Old Orders for New: Ecology, Animal Rights, and the Poverty of Humanism," *Diacritics* 28, 2 (summer 1998): 21–40.

17. Jean-François Lyotard, *The Inhuman: Reflections on Time*, trans. Geoffrey Bennington and Rachel Bowlby (Stanford: Stanford University Press, 1991), 4.

18. Slavoj Žižek, in his critique of liberal democracy, articulates this linkage even more pointedly (though not entirely unproblematically). In *Looking Awry: An Introduction to Jacques Lacan through Popular Culture* (Cambridge: MIT Press, 1991), Žižek writes that "the subject of democracy is thus a pure singularity, emptied of all content, freed from all substantial ties" but that "the problem with this subject does not lie where neoconservatism sees it." It is not that "this abstraction proper to democracy dissolves all concrete substantial ties, but rather that *it can never dissolve them.*" The subject of democracy is thus "smeared with a certain 'pathological' stain" (to use Kant's term) (164–65). In *Tarrying with the Negative: Kant, Hegel, and the Critique of Ideology* (Durham, N. C.: Duke University Press, 1993), Žižek elaborates the linkage between the "abstract" subject of liberalism and the unfortunate term "political correctness" even more specifically by arguing that in "the unending effort to unearth traces of sexism and racism in oneself," in fact, "the PC type is not ready to renounce what really matters: 'I'm prepared to sacrifice everything *but that*'—but what? The very gesture of self-sacrifice." Thus, "In the very act of emptying the white- male-heterosexual position of all positive content, the PC attitude retains it as a universal form of subjectivity" (213–14).

19. Jacques Derrida, "The Animal That Therefore I Am (More to Follow)," trans. David Wills, 73 (unpublished manuscript; see chap. 2, n. 22 below).

Chapter 1

1. Luc Ferry, *The New Ecological Order*, trans. Carol Volk (Chicago: University of Chicago Press, 1995), xx. Further references are given in the text.

2. Another important liberal European intellectual—Niklas Luhmann—comes to mind here. Though working in an explicitly posthumanist framework (systems theory), Luhmann is also bothered by the Greens for these reasons, as he makes clear in *Ecological Communication*, trans. John Bednarz Jr. (Chicago: University of Chicago Press, 1989). For a discussion of the dangers of ethics from Luhmann's point of view, see William Rasch's informative discussion in "Immanent Systems, Transcendental Temptations, and the Limits of Ethics," *Cultural Critique* 30 (spring 1995), esp. 213 ff.

3. Fredric Jameson, *Late Marxism: Adorno, or The Persistence of the Dialectic* (London: Verso, 1990), 249; emphasis mine.

4. An example of Ferry's superficial engagement with postmodern theory is evinced in his rather remarkable misreading of Félix Guattari's interest in ecology, where he mistakes what Guattari calls "resingularization" for conservative identity politics. Anyone who has read any of Guattari's work over the past thirty years knows that he has never held that identity is a positivity that can be either accomplished or restored in the sense Ferry attributes to him. The same should be said, of course, for Ferry's claim that Guattari's contention that "there is no reason to ask immigrants to give up *their national affiliation or the cultural traits that cling to their being*" is "a 'leftist' version of racism" (114). This rather absurd charge might be plausible were it not that Guattari's work long ago made it clear that he does not believe in the existence of races, or of their equivalent in terms of cultural identity.

5. Michel Foucault, "Truth and Power," in *The Foucault Reader*, ed. Paul Rabinow (New York: Pantheon, 1984), 58.

6. Gregory Bateson, "Form, Substance, and Difference," in his *Steps to an Ecology of Mind* (New York: Ballantine, 1972), 451. Further references are given in the text.

7. See Kenneth Burke's discussion of technology in the postscript to the second edition of his *Attitudes toward History* (Berkeley: University of California Press, 1984), 396. Rifkin, as is well known, is one of the more socially visible critics of the current plunge into genetic engineering, especially of animals. Heidegger's aversion to the domination of technology is perhaps the single most famous instance in modern philosophy and will be elaborated in the next chapter.

8. Tim Luke, "The Dreams of Deep Ecology," *Telos* 76 (summer 1988): 51. Further references are given in the text.

9. Chantal Mouffe, *The Return of the Political* (London: Verso, 1993), 10.

10. Such tried if not quite true rhetorical maneuvers will be familiar to readers who have encountered them in other liberal intellectuals such as Richard Rorty, who—despite his substantially more sophisticated and productive engagement with postliberal, postmodern theorists—sometimes stoops to paint people interested in, say, Marxism as nothing more than religious fanatics bent on salvation rather than on reasoned, constructive change in society's material structures and maldistribution of wealth and power. See, for example, Rorty's characterization of opponents of

liberalism as "people who have always hoped to become a New Being, who have hoped to be converted rather than persuaded," in his *Objectivity, Relativism, and Truth* (Cambridge: Cambridge University Press, 1991), 29. Further references are given in the text.

11. Arran E. Gare, *Postmodernism and the Environmental Crisis* (New York: Routledge, 1995), 77–78.

12. Fredric Jameson, *The Seeds of Time* (New York: Columbia University Press, 1994), xii. Further references are given in the text.

13. Tom Regan, *The Case for Animal Rights* (Berkeley: University of California Press, 1983), 277.

14. Peter Singer, "Prologue: Ethics and the New Animal Liberation Movement," in *In Defense of Animals*, ed. Peter Singer (New York: Harper and Row, 1985), 5. Further references are given in the text.

15. See Peter Singer's *Animal Liberation* (New York: Avon, 1975), 21–22, and Regan's *Case for Animal Rights*, 324.

16. See Mary Midgley, "The Significance of Species," in *The Animal Rights/ Environmental Ethics Debate: The Environmental Perspective*, ed. Eugene C. Hargrove (Albany: SUNY Press, 1992), 121–36.

17. Stephan Zak, "Ethics and Animals," *Atlantic Monthly*, March 1989, 71. Further references are given in the text. See also Mary Midgley, "Persons and Nonpersons," in Singer, *In Defense of Animals*.

18. Deborah Slicer, "Your Daughter or Your Dog? A Feminist Assessment of the Animal Research Issue," *Hypatia* 6, 1 (spring 1991): 110. Further references are given in the text.

19. Tom Regan, "The Case for Animal Rights," in Singer, *In Defense of Animals*, 22.

20. Ferry attempts to salvage this point by holding to an impossible distinction between "a simple factual *situation*, even if is it intangible like belonging to one of the two sexes" (!) and "a *determination* which in some sense shapes us outside of all voluntary activity" (115). Here as throughout, Ferry is desperate to maintain as differences in *kind* what can only be defended as differences in *degree*.

21. In fairness, Regan points out that more sophisticated forms of contractarianism, such as John Rawls's *A Theory of Justice* attempt to force "contractors to ignore the accidental features of being a human being" to meet this problem—but only at the price of denying any direct duties to "those human beings who do not have sense of justice—young children, for instance, and many mentally retarded humans" (17).

22. See Cary Wolfe, *Critical Environments: Postmodern Theory and the Pragmatics of the "Outside"* (Minneapolis: University of Minnesota Press, 1998).

23. See Marc Bekoff and Dale Jamieson, eds., *Interpretation and Explanation in the Study of Animal Behavior*, vol. 1 (Boulder, Colo.: Westview Press, 1990); Marian Stamp Dawkins, *Through Our Eyes Only? The Search for Animal Consciousness* (New York: W. H. Freeman/Spektrum, 1993); Donald Griffin, *Animal Minds* (Chicago: University of Chicago Press, 1992); and Paola Cavalieri and Peter Singer, eds., *The Great Ape Project: Equality beyond Humanity* (New York: St. Martin's Press, 1993).

24. Richard Ryder, "Sentientism," in Cavalieri and Singer, *Great Ape Project*, 220.

25. Jane Goodall, "Chimpanzees—Bridging the Gap," in Cavalieri and Singer, *Great Ape Project*, 12.

26. See Georges Bataille, *Theory of Religion*, trans. Robert Hurley (New York: Zone Books, 1992), esp. 17–61; Jacques Derrida, "'Eating Well,' or The Calculation of the Subject: An Interview with Jacques Derrida," in *Who Comes after the Subject?* ed. Eduardo Cadava, Peter Connor, and Jean-Luc Nancy (New York: Routledge, 1991), 96–119; Jacques Derrida, "Force of Law: The 'Mystical Foundation of Authority,'" *Cardozo Law Review* 11, 919 (1990), 951, 953. For a fuller discussion of these issues, see chapter 3 below.

27. Slavoj Žižek, *Looking Awry: An Introduction to Jacques Lacan through Popular Culture* (Cambridge: MIT Press, 1992), 26.

28. Slavoj Žižek, *Enjoy Your Symptom! Jacques Lacan in Hollywood and Out* (New York: Routledge, 1992), 181. Further references are given in the text.

Chapter 2

1. *The Wittgenstein Reader*, ed. Anthony Kenny (Oxford: Blackwell, 1994), 213. Further references are given in the text.

2. Quoted in Vicki Hearne, *Adam's Task: Calling Animals by Name* (New York: Random House, 1987), 4. Further references are given in the text.

3. Vicki Hearne, *Animal Happiness* (New York: HarperCollins, 1994), 167. Further references are given in the text.

4. Stanley Cavell, *Philosophical Passages: Wittgenstein, Emerson, Austin, Derrida* (Oxford: Blackwell, 1995), 151–52. Further references are given in the text.

5. Stanley Cavell, *The Claim of Reason: Wittgenstein, Skepticism, Morality, and Tragedy* (Oxford: Oxford University Press, 1979), 187–88.

6. See Cavell's *Conditions Handsome and Unhandsome: The Constitution of Emersonian Perfectionism* (Chicago: University of Chicago Press, 1990), 1–33, 101–26.

7. Tom Regan, "The Case for Animal Rights," in *In Defense of Animals*, ed. Peter Singer (New York: Harper and Row, 1985), 16. Further references are given in the text. See here especially Regan's detailed discussion of Rawls's contract theory and Kant's "indirect duty" view in his *The Case for Animal Rights* (Berkeley: University of California Press, 1983), 163–94.

8. Stanley Cavell, *This New yet Unapproachable America: Lectures after Emerson after Wittgenstein* (Albuquerque, N.Mex.: Living Batch Press, 1989), 41–42. Further references are given in the text.

9. Here Cavell's reading of the human form of life in Wittgenstein links up directly with his rendering of Emersonian "perfectionism." See his introduction to *Conditions Handsome and Unhandsome*.

10. See Daniel Dennett, *Consciousness Explained* (Boston: Little, Brown, 1991), esp. 431 ff. Further references are given in the text.

11. Stanley Cavell, *In Quest of the Ordinary: Lines of Skepticism and Romanticism* (Chicago: University of Chicago Press, 1988), 31. Further references are given in the text.

12. What Wittgenstein means by the term "language game," Lyotard writes, "is that each of the various categories of utterance can be defined in terms of rules specifying their properties and the uses to which they can be put—in exactly the same way as the game of chess is defined by a set of rules determining the properties of each of the pieces, in other words, the proper way to move them." Jean-François Lyotard, *The Postmodern Condition: A Report on Knowledge*, trans. Geoff Bennington and

Brian Massumi, foreword by Fredric Jameson (Minneapolis: University of Minnesota Press, 1984), 10. Further references are given in the text. As for the "grand
metanarratives," Lyotard writes, "The sometimes violent divergences between political liberalism, economic liberalism, Marxism anarchism, the radicalism of the
Third Republic and socialism, count for little next to the abiding unanimity about
the end to be attained. The promise of freedom is for everyone the horizon of progress and its legitimation. . . .[Yet] it was not a lack of progress but, on the contrary,
development (technoscientific. artistic, economic, political) that created the possibility of total war, totalitarianisms, the growing gap between the wealth of the North
and the impoverished South, unemployment and the 'new poor,' general decultura-
tion and the crisis in education (in the transmission of knowledge), and the isolation
of the artistic avant-gardes" (*The Postmodern Explained*, ed. Julian Pefanis and Morgan Thomas, trans. Julian Pefanis, Morgan Thomas, et al., afterword by Wlad
Godzich [Minneapolis: University of Minnesota Press, 1993], 82). Further references are given in the text.

13. Jean-François Lyotard, *The Inhuman*, trans. Geoffrey Bennington and Rachel
Bowlby (Stanford: Stanford University Press, 1991). 2–3. Further references are
given in the text.

14. As Lyotard notes, the idea that that "nothingness" could be filled in or is
simply epiphenomenal—even if we remain squarely within formalism or conventionalism—founders on the aporia that Russell attempts to arrest with the theory of
logical types (*The Differend: Phrases in Dispute* [Minneapolis: University of Minnesota Press, 1988], 138). This is a topic I have taken up elsewhere on the work on
Niklas Luhmann. See Cary Wolfe, *Critical Environments: Postmodern Theory and the
Pragmatics of the "Outside"* (Minneapolis: University of Minnesota Press, 1998),
65–70, 117–28.

15. Here we might consult, among many others, Diana Fuss in her editorial introduction to the collection *Human, All Too Human* (New York: Routledge, 1996),
which points out that "the vigilance with which the demarcations between humans
and animals, humans and things, and humans and children are watched over and
safeguarded tells us much about the assailability of what they seek to preserve: an abstract notion of the human as unified, autonomous, and unmodified subject,"
whereas the "all too" of Nietzsche's famous formulation "all too human" "locates at
the center of the human some unnamed surplus—some residue, overabundance, or
excess" (Lyotard's "remainder")—that is "embedded inside the human as its condition of possibility" (3–4).

16. Jean-François Lyotard and Jean-Loup Thébaud, *Just Gaming*, trans. Wlad
Godzich, afterword by Samuel Weber (Minneapolis: University of Minnesota Press,
1985), 41–42. Further references are given in the text.

17. And it connects rather directly, as Simon Critchley has noted, with Cavell's sense
of the ethical import of skepticism. As Critchley writes, "In Stanley Cavell's terms, it is
the very unknowability of the other, the irrefutability of scepticism, that initiates a relation to the other based on acknowledgement and respect. The other person stands in a
relation to me that exceeds my cognitive powers, placing me in question and calling me
to justify myself." See "Deconstruction and Pragmatism—Is Derrida a Private Ironist
or a Public Liberal?" in *Deconstruction and Pragmatism*, ed. Chantal Mouffe (London:
Routledge, 1996), 32. Further references are given in the text.

18. In fairness, Lyotard is quick to specify his difference with Lévinas late in *Just Gaming*, when he write that in Lévinas's view "it is the transcendental character of the other in the prescriptive relation, in the pragmatics of prescription, that is, in the (barely) lived experience of obligation, that is truth itself. This 'truth' is not ontological truth, it is ethical. But it is a truth in Lévinas' own terms. Whereas, for me, it cannot be the truth. . . . It is not a matter of privileging a language game above others" but rather of "the acceptance of the fact that one can play several games" (*Just Gaming*, 60–61). Now this may remove Lyotard somewhat from the sort of objection readily raised against Lévinas's position—indeed it is raised by Lyotard himself in his "Lévinas notice" in *The Differend* under the guise of "the commentator," who would object that "the less I understand you, he or she says to the Lévinassian (or divine) text, the more I will obey you by that fact; for, if I want to understand you (in your turn) as a request, then I should not understand you as sense" (115). But that reservation toward Lévinas does not remove Lyotard's sense of ethics from the paradoxical problem noted by several critics, including Samuel Weber in his afterword to *Just Gaming:* if "it is necessary for a singular justice to impose its rule on all the other games, in order that they may retain their own singularity," then it is also "necessary to be able to distinguish between this violence, in some way legitimate and necessary, and 'terror,' described as the attempt to reduce the multiplicity of the games or players through exclusion or domination. But how, then, can we conceive of such a justice, one that assures, 'by a prescriptive of universal value,' the *nonuniversality* of singular and incommensurable games?" (103).

19. John Llewelyn, "Am I Obsessed by Bobby? (Humanism of the Other Animal)," in *Re-reading Levinas*, ed. Robert Bernasconi and Simon Critchley (Bloomington: Indiana University Press, 1991), 235. Further references are given in the text.

20. And here he seems to contest the reading of Kant given by Tom Regan in *The Case for Animal Rights*. On this point Llewelyn writes, "It is argued that Kant's concession that we have indirect duties to animals can be reduced to absurdity on the grounds that rationality is the only morally relevant characteristic that he can admit by which to distinguish animals from other nonhuman beings and that therefore, if we are to refrain from treating animals only as means because that is likely to lead us to treat fellow humans as means only, we should for the same reason refrain from treating only as means inanimate objects like hammers" (240).

21. Interestingly enough, Lyotard suggests in passing that he wants to maintain *in principle* the possibility of nonhuman animals as part of this community of reasonable beings, for as he states in *The Differend*, "The community of practical, reasonable beings (obligees and legislators, since that is the hypothesis) includes just as well entities that would not be human. This community cannot be empirically tested. Concession: we can't really say if and how the object or referent intended by the Idea of this community is possible, but it is at least possible to conceptualize this community, it is not a 'being of reason,' or an empty concept: it is a community of persons. . . .On the scale of the single entity, it signifies autonomy. The community of practical, reasonable beings merely extends this principle of autonomy onto the scale of all possible entities, on the condition that they satisfy the definition of a practical, reasonable being, that is, of a person" (126). Theoretically, on this view, *if it could be shown* that some animals fulfill the definition of a practical, reasonable being

in the Kantian sense, then they would presumably fall under the sphere of ethical consideration. But if that community "cannot be empirically tested," and since in Kant the bar of definition is set in such a way that it coincides more or less in fact with the subject qua human as that which can "universalize its maxim," then we are forced to say that Lyotard's Kantianism excludes the animal other, if not on principle, then certainly in effect. Hence the distinction between species doesn't necessarily do any work in Lyotard's reading of Kantian ethics; but then, it doesn't need to. See also here Steve Baker's interesting discussion of Lyotard's contention, in *Signé Malraux* (Paris: Grasset, 1996), that cats exist "at thresholds we do not see, where they sniff some 'present beyond,'" and in doing so live a "questioning" existence that is particularly instructive for the writer and the philosopher (Steve Baker, *The Postmodern Animal* [London: Reaktion Books, 2000], 184).

22. Jacques Derrida, *Of Spirit: Heidegger and the Question*, trans. Geoffrey Bennington and Rachel Bowlby (Chicago: University of Chicago Press, 1989), 11. Further references are given in the text. See Derrida's own partial list of his texts in which the animal has appeared, in "The Animal That Therefore I Am (More to Follow)," 54–59, which has appeared in French in the volume of essays that grew out of the conference, titled *L'animal autobiographique*, ed. Marie-Louise Mallet (Paris: Galilée, 1998). My page citation refers to the text of David Wills's superb English translation, which is as yet unpublished.

23. Jacques Derrida, *"Geschlecht* II: Heidegger's Hand," trans. John P. Leavey Jr., in *Deconstruction and Philosophy*, ed. John Sallis (Chicago: University of Chicago Press, 1986), 173. Further references are given in the text.

24. See also *Of Spirit*, 56, where Derrida writes: "I do not mean to criticize this humanist teleology. It is no doubt more urgent to recall that, in spite of all the denegations or all the avoidances one could wish, it has remained *up till now* (in Heidegger's time and situation, but this has not radically changed today) the price to be paid in the ethico-political denunciation of biologism, racism, naturalism, etc. . . . Can one transform this program? I do not know." The recent work from Cerisy suggests, however, that Derrida will die trying to theorize just such a transformation and that, in truth, he has all along been engaged in just such a project—hence this statement is perhaps too modest.

25. On the point of technology, see especially the discussion of Heidegger's opposition of handwriting and the typewriter in Derrida's *"Geschlecht* II," 178–81, which condenses many of these themes. As he puts it, for Heidegger "typographic mechanization destroys this unity of the world, this integral identity, this proper integrity of the spoken word that writing manuscripts, at once because it appears closer to the voice or body proper and because it ties together the letters, conserves and gathers together" (178). It is thus "a-signifying" because "it loses the hand," hence, as Heidegger puts it, "In typewriting, all men resemble one another" (179). "The protest against the typewriter," Derrida notes, "also belongs—this is a matter of course—to an interpretation of technology [*technique*], to an interpretation of politics starting from technology," but also and more importantly to a "devaluation of writing in general" as "the increasing destruction of the word or of speech" in which "the typewriter is only a modern aggravation of the evil" (180).

26. For Derrida, this "vulnerability" and "passivity" connect very directly to the question of shame and the motif of nakedness before the gaze of the other that struc-

tures the entire essay. In what sense can one be naked—and perhaps naked as before no *other* other—before the gaze of an animal? "I often ask myself," he writes, "just to see, *who I am* (following) at the moment when, caught naked, in silence, by the gaze of an animal, for example the eyes of a cat, I have trouble, yes, a bad time, overcoming my embarrassment" (5). In a sense, this means nothing more than the fact that Derrida sees *himself* as a philosopher, for as he notes, Descartes, Kant, Heidegger, Lacan, and Lévinas produce discourses that are "sound and profound, but everything goes on as if they themselves had never been looked at, and especially not naked, by an animal that addressed them. At least everything goes on as though this troubling experience had not been theoretically registered, supposing they had experienced it at all, at the precise moment"—and here we recall Heidegger's use of the form of the *thesis*—"when they made of the animal a *theorem*" (20). Derrida, on the other hand, wants to insist on the "unsubstitutable singularity" of the animal (in this case "a real cat") and suggests that our readiness to turn it into a "theorem" is at base a panicked horror at our own vulnerability, our own passivity—in the end, our own mortality. "As with every bottomless gaze," he writes, "as with the eyes of the other, the gaze called 'animal' offers to my sight the abyssal limit of the human: the inhuman or the ahuman, the ends of man. . . . And in these moments of nakedness, under the gaze of the animal, everything can happen to me, I am like a child ready for the apocalypse" (18).

27. Richard Beardsworth, *Derrida and the Political* (London: Routledge, 1996), 129. Further references are given in the text.

28. It should be noted here that, for Singer's own part, there appears to be no love lost either. When asked recently about the relevance of theory associated with "postmodernism" to bioethics, he replied, "Life's too short for that sort of thing." See Jeff Sharlet, "Why Are We Afraid of Peter Singer?" *Chronicle of Higher Education* 46, 27 (March 10, 2000): A22.

29. See, for example, Derrida, "*Geschlecht* II" and "*Geschlecht*: Sexual Difference, Ontological Difference," trans. Ruben Berezdivin, in *A Derrida Reader: Between the Blinds*, ed. Peggy Kamuf (New York: Columbia University Press, 1991), 380–402. For Lévinas, see Jacques Derrida, "At This Very Moment in This Work Here I Am," in *Re-reading Levinas*, 11–48, and the selection from "Choreographies," trans. Christie V. McDonald, in *Derrida Reader*, 440–56. Further references to these works are given in the text.

30. See Derrida's "Choreographies," 450–51, and also "At This Very Moment," 40–44.

31. Quoted in Vicki Kirby, *Telling Flesh: The Substance of the Corporeal* (New York: Routledge, 1997), 90. Further references to Kirby's book are given in the text.

32. This essay is also part of the material delivered by Derrida at Cerisy-la-Salle in 1997, but unlike "The Animal That Therefore I Am," it has not appeared in either English or French. I am working from the manuscript of the translation by David Wills that will appear in *Zoontologies: The Question of the Animal*, ed. Cary Wolfe (Minneapolis: University of Minnesota Press, 2002). Page references are given parenthetically.

33. J. T. Fraser, *Of Time, Passion, and Knowledge: Reflections on the Strategy of Existence* (New York: George Braziller, 1975). I am drawing on the popularization given in Jeremy Rifkin's *Algeny* (New York: Penguin, 1984), 186–91.

34. Eva M. Knodt, foreword to Niklas Luhmann, *Social Systems*, trans. John Bednarz Jr. with Dirk Baecker (Stanford: Stanford University Press, 1995), xxxi. Further references are given in the text.

35. Francisco J. Varela, "The Reenchantment of the Concrete," in *Incorporations*, ed. Jonathan Crary and Sanford Kwinter (New York: Zone Books, 1992), 320.

36. *Of Grammatology*, trans. Gayatri Chakravorty Spivak (Baltimore: Johns Hopkins University Press, 1976), 70. See also, for example, *Limited Inc.*, ed. Gerald Graff (Evanston, Ill.: Northwestern University Press, 1988), where Derrida distinguishes his view of the relation between scientific theories and theories of language from those of John Searle (118, 69–70).

37. Specifically, in Wolfe, *Critical Environments*, 78–84.

38. But for useful overviews of this material see, for example, Paola Cavalieri and Peter Singer, eds., *The Great Ape Project: Equality beyond Humanity* (New York: St. Martin's Press, 1993), and Robert W. Mitchell, Nicholas S. Thompson, and H. Lyn Miles, eds., *Anthropomorphism, Anecdotes, and Animals* (Albany: SUNY Press, 1997).

39. Humberto Maturana and Francisco Varela, *The Tree of Knowledge: The Biological Roots of Human Understanding*, rev. ed., trans. Robert Paolucci, foreword by J. Z. Young (Boston: Shambhala Press, 1992), 165. Further references are given in the text.

40. This is true even in animals with the most minimal cephalization, such as the social insects, whose third-order couplings are, however, markedly rigid and inflexible because of the limits placed on the possible concentration of nervous tissue by their hard exteriors of chitin (188). Hence their plasticity is limited and their individual ontogenies are of little importance in explaining their behavior, even though we cannot understand their behavior without understanding their broadly shared ontogenies.

41. This is why, according to Maturana and Varela, the "language of bees" is not a language; it is a largely fixed system of interactions "whose stability depends on the genetic stability of the species and not on the cultural stability of the social system in which they take place" (208).

42. The example they give is of the chimp Lucy, who, on the verge of a tantrum on seeing her human "parents" about to leave, turned to her keepers and signed in Ameslan, "Lucy cry"—a "linguistic distinction of an action performed" (215).

43. Here one might readily think of the example of animals used in factory farming, but also—on the other, human, hand—of the example of the "wolf children" cited by Maturana and Varela, two Hindu girls who were raised by a pack of wolves, without human contact, whose behaviors (modes of ambulation, dietary preferences, signifying repertoires, and so on) were in all significant respects canid and not human (*Tree*, 128–30).

44. Gregory Bateson, "A Theory of Play and Fantasy," in his *Steps to an Ecology of Mind* (New York: Ballantine, 1972), 179. Further references are given in the text.

45. It is significant in this regard—though not at all surprising—that Maturana and Varela are therefore willing to grant the existence of "cultural behaviors" in nonhuman social groups (*Tree*, 194–201).

46. See Wolfe, *Critical Environments*, 57 ff., on Bateson and, on representationalism, xi–xxiv, 12–22, and 41–71. On Dennett, see Richard Rorty's critique of Dennett's view in *Consciousness Explained* that it is possible to construct an "objective"

"heterophenomenological text," in the same way that it is possible to provide a "correct" interpretation of a literary text. As Rorty puts it, "no up-to-date practitioner of hermeneutics—the sort who agrees with Derrida that there is no transcendental signified and with Gadamer that all readings are prejudiced—would be caught dead talking about the 'right interpretation'" (Richard Rorty, "Comments on Dennett," *Synthese* 53 [1982]: 184).

47. Dietrich Schwanitz, "Systems Theory according to Niklas Luhmann—Its Environment and Conceptual Strategies," *Cultural Critique* 30 (spring 1995): 156. Further references are given in the text.

48. Humberto R. Maturana, "Science and Daily Life: The Ontology of Scientific Explanations," in *Research and Reflexivity*, ed. Frederick Steier (London: Sage, 1991), 34. Further references are given in the text.

49. Rodolphe Gasché, *Inventions of Difference: On Jacques Derrida* (Cambridge: Harvard University Press, 1994), 4.

50. There is a difference of accent here, in other words, and the seriousness of that difference is of some moment and rests in no small part on whether one shares this characterization of Derrida's view. As Schwanitz points out, in comparison with systems theory, "Derrida reverses the relation between disorder and order. According to him, the level of order consists of the text of Western metaphysics that is brought about by a fundamental attribution of meaning to the simultaneity of the idea and the use of signs. In terms of systems theory, constative language is a kind of self-simplification of writing for the benefit of logos. On the other hand, writing as the basic differentiation within the use of signs that is also inherent in the spoken word, undertakes a permanent renewal of complexity and contingency through dissemination and dispersion, which in turn is again reduced by logocentric self-simplification. According to Luhmann, however, the paradox of self-referentiality comes first and the asymmetry produced by temporalization comes second. The opposite is true for Derrida. The 'illegitimate' asymmetry as a form of domination comes first and is then dissolved in the paradox of time" (155).

51. Carolyn Merchant, *Radical Ecology* (New York: Routledge, 1993), 107.

52. Niklas Luhmann, "Deconstruction as Second-Order Observing," *New Literary History* 24 (1993): 770.

53. Niklas Luhmann, *Observations on Modernity*, trans. William Whobrey (Stanford: Stanford University Press, 1998), 108. Further references are given in the text.

54. Humberto R. Maturana and Francisco J. Varela, *Autopoiesis and Cognition: The Realization of the Living* (Dordrecht: Reidel, 1980), 39.

Chapter 3

1. The film has generated a good deal of thoughtful criticism, both academic and journalistic, focusing on its complex messages about gender and sexuality. See, for example, Douglas Crimp, "Right on, Girlfriend," in *Fear of a Queer Planet: Queer Politics and Social Theory*, ed. Michael Warner (Minneapolis: University of Minnesota Press, 1993), 300–320; and Diana Fuss, "Monsters of Perversion: Jeffrey Dahmer and *The Silence of the Lambs*," in *Media Spectacles*, ed. Marjorie Garber, Jann Matlock, and Rebecca L. Walkowitz (New York: Routledge, 1993), 181–207. A cogent and ambitious attempt to see the discourses of gender and sexuality operating asymmetrically in the film can be found in Elizabeth Young, "*The Silence of the Lambs* and the

Flaying of Feminist Theory," *Camera Obscura* 27 (September 1991): 5–35. See also Judith Halberstam, "Skin-Flick: Posthuman Gender in Jonathan Demme's *The Silence of the Lambs*," *Camera Obscura* 27 (September 1991): 37–52. A reading of the film that emphasizes class rather than gender and sexuality is Adrienne Donald, "Working for Oneself: Labor and Love in *The Silence of the Lambs*," *Michigan Quarterly Review* 31, 3 (summer 1992): 347–60.

2. See Carol Clover, *Men, Women, and Chainsaws: Gender in the Modern Horror Film* (Princeton: Princeton University Press, 1992).

3. Fredric Jameson, "Reification and Utopia in Mass Culture," *Social Text* 1 (winter 1979): 147–48.

4. Stephen King, *Danse Macabre* (New York: Berkeley Books, 1981), 31.

5. See Clover, *Men, Women, and Chainsaws*, 223 n. 132.

6. And here we should remark that we will treat only the film, which in this particular diverges crucially from Thomas Harris's novel.

7. Judith Butler, *Bodies That Matter: On the Discursive Limits of "Sex"* (New York: Routledge, 1993), 18.

8. Jacques Derrida, "'Eating Well,' or The Calculation of the Subject: An Interview with Jacques Derrida," in *Who Comes after the Subject?* ed. Eduardo Cadava, Peter Connor, and Jean-Luc Nancy (New York: Routledge, 1991), 114. Further references are given in the text.

9. Jacques Derrida, "Force of Law: The 'Mystical Foundation of Authority," *Cardozo Law Review* 11, 919 (1990): 951, 953. Further references are given in the text.

10. For a remarkable contemporary example of the uncanny, one that creates its effects by exploiting the logic linking pets, hostages, and sacrifice, see Lisa Tuttle's story "Replacements," in *Metahorror*, ed. Dennis Etchison (New York: Dell, 1992), 73–93.

11. Deleuze and Guattari, in their insightful meditation on "becoming-animal" in *A Thousand Plateaus*, will help us make the point. "We must distinguish between three kinds of animals," they write in a passage worth quoting at length. "First, individuated animals, family pets, sentimental, Oedipal animals each with its own petty history, 'my' cat, 'my' dog. These animals invite us to regress, draw us into narcissistic contemplation, and they are the only kind of animal psychoanalysis understands. And then there is a second kind: animals with characteristics or attributes; genus, classification or State animals; animals as they are treated in the great divine myths, in such a way as to extract from them series or structures, archetypes or models (Jung is in any event profounder than Freud). Finally, there are more demonic animals, pack or affect animals that form a multiplicity, a becoming, a population." See Gilles Deleuze and Félix Guattari, *A Thousand Plateaus: Capitalism and Schizophrenia*, trans. Brian Massumi (Minneapolis: University of Minnesota Press, 1987), 241.

12. See Karen Warren, "The Power and Promise of Ecological Feminism," *Environmental Ethics* 12 (summer 1990): 125–46.

13. Carol J. Adams, "Ecofeminism and Eating of Animals," *Hypatia* 6, 1 (spring 1991): 136–37. See also Adams's chapter "The Rape of Animals, the Butchering of Women," in her *The Sexual Politics of Meat* (New York: Continuum, 1990).

14. Donna J. Haraway, "Situated Knowledges: The Science Question in Feminism and the Privilege of Partial Perspective," in her *Simians, Cyborgs, and Women:*

The Reinvention of Nature (New York: Routledge, 1991), 188. Further references are given in the text. In "Monsters of Perversion," Fuss is very instructive on the film's cutting techniques, pointing out in particular that whatever Bill dishes out in the way of objectification and scoping, he receives in return: there are "more than ten extreme closeups [that] fetishize different parts of Gumb's body: eyes, hands, neck, nipples, and lips. . . .Gumb is optically dismembered as savagely as he is known to have mutilated his victims" (194).

15. For an elaborate version of this argument, see Noel Carroll, *The Philosophy of Horror, or Paradoxes of the Heart* (New York: Routledge, 1990). Halberstam also has some observations along these lines in "Skin-Flick."

16. As Franco Moretti has pointed out, in the genre of detection (which this film draws on freely), it is the function of the detective to gradually individualize the criminal, so that guilt may be displaced from the collective social body onto the individualized body of the perverse or "animalistic" criminal. (In this light, the genius of Poe's use of the orangutan as the killer in "The Murders in the Rue Morgue" perhaps needs little amplification.) As long as the criminal is unknown, guilt may be generalized: all of society is potentially guilty. The role of the detective as analyst, then, is to restore rationality and harmony to liberal humanist society by reaffirming once again that societies don't commit crimes, only criminals do. See Moretti's essay "Clues," in his *Signs Taken for Wonders*, trans. Susan Fischer, David Forgacs, and David Miller (London: New Left Books, 1988).

17. Slavoj Žižek, *Enjoy Your Symptom! Jacques Lacan in Hollywood and Out* (New York: Routledge, 1992), 136. This work is cited in the text as *Enjoy*. Žižek does, in fact, briefly address *The Silence of the Lambs* in his later work, *Tarrying with the Negative: Kant, Hegel, and the Critique of Ideology* (Durham, N. C.: Duke University Press, 1993), 47–49, 106, an account we discovered only after our essay was written. We are pleased to find several parallels between Žižek's use of himself and our own. Žižek, too, focuses on the sadomasochism at the core of the psychoanalytic relationship between Lecter and Starling. This much is clear when he writes that the analysand's "fullness of being" is a kind of "stuff" that the analyst pulverizes and "swallows." But Žižek's Heideggerian language here, while noting the coimplication of analysis and sadomasochism, stops short of our insistence on the materiality of symbolic substitution. That is, Lecter surely says in all but words, as does Žižek, "'Mange ton *Dasein!*'—'Eat your being-there!'" But our point, what makes this horrible and not just silly, is that Lecter pursues to its logical terminus the practice of meat eating and sacrifice. He eats the *body*, and not just the *Dasein*, of the other of whatever species. This, it seems to us, is the very point of Derrida's most recent work on animality, which we have already quoted above.

18. Freud wrote this to his friend Wilhelm Fliess in 1897. See *The Complete Letters of Sigmund Freud to Wilhelm Fliess, 1887–1904*, ed. and trans. Jeffrey Moussaieff Masson (Cambridge: Belknap Press of Harvard University Press, 1985), 279.

19. Sigmund Freud, *Civilization and Its Discontents*, trans. and ed. James Strachey (New York: Norton, 1961), 59, 52. Further references are given in the text.

20. Apropos of the mention of Ridley Scott, it should be noted that we will not be remarking on the disappointing sequel to *The Silence of the Lambs*, the film *Hannibal* (2001), which Scott directed, primarily because it adds little if anything to the motifs and problematics introduced in *Silence* and indeed pushes them to

disappointingly Hollywoodized, predictable, and (if you will allow the expression) ham-fisted results (e.g., Lecter's parting kiss to Starling, his self-dismemberment/ symbolic castration in the same scene, and perhaps most comically, his culinary preparation and feeding to its owner of the exposed brain of the character played by Ray Liotta, whose on-screen frat-boy persona only makes the entire scene seem all the more contrived and unintentionally comedic and prankish).

21. Henry Sussman, *Psyche and Text: The Sublime and the Grandiose in Literature, Psychopathology, and Culture* (New York: SUNY Press, 1993), 182. Further references are given in the text.

22. Georges Bataille, *Theory of Religion*, trans. Robert Hurley (New York: Zone, 1989), 18. Further references are given in the text.

23. We should not make the mistake of thinking that the scream qua object here belongs to Bill's victims, for to do so would reinscribe a sexist reading on the body of the woman whose scream of enjoyment is at her own violation and death.

24. Philippe Lacoue-Labarthe, "Typography," in his *Typography: Mimesis, Philosophy, Politics* (Cambridge: Harvard University Press, 1989), 116. Lacoue-Labarthe's monumental essay explicitly engages, and critiques, Girard's theories of sacrifice and mimetic contagion on pages 102–19. For Girard's own version of these ideas, see— among other works—René Girard, *Violence and the Sacred*, trans. Patrick Gregory (Baltimore: Johns Hopkins University Press, 1977). Further references are given in the text.

25. Herbert Marcuse, "On the Affirmative Character of Culture," in his *Negations* (Boston: Beacon Press, 1968), 88–133.

26. The elitism of Lecter is visible not least of all in the character's many allusions to Conan Doyle's master detective Sherlock Holmes—and beyond that, to Poe's Auguste Dupin and even his Legrand: the love of classical music (Holmes), the seeming ability to mind-read, to reproduce the state of mind of the criminal, and indeed to solve the entire case without ever setting foot on the street (Dupin), the essential bohemianism and connoisseurship of all things fine (Holmes), the surprising and explosive physical strength (Holmes), the mastery of disguise and total control of the body (Holmes again), and the indulgence in anagrams and cryptography (Legrand). Lecter is not the anti-Holmes, however, as we might suspect, but rather the ultra-Holmes, the Holmes of cocaine addiction, solitary violin playing, depression, and dangerously antisocial bohemian tendencies, the Holmes who looks like a great bird with pointed beak—in short, Holmes the gothic monster, who, to borrow Žižek's phrase, "probes too deeply into obscure origins." Lecter completes the trajectory only hinted at in Holmes; he is Holmes beyond Watson, Oedipal father beyond incest taboo, meat eater beyond speciesism.

27. For a cogent reading of the way the ending of the film presents the lines of class and consumerism and how this constitutes a "final postmodern joke upon the film's audience"—a view to which we give a somewhat different spin—see Young, 26–27. Young also points to the surfacing of a race discourse in the final scene, arguing that, despite Demme's own politically left views on, and interest in, Afro-Caribbean culture (he made *Haiti Dreams of Democracy* in 1987), the final credits sequence amounts to a "collapse into that most tired of racist Hollywood conventions, the scenario of fully realized white characters set against an undifferentiated backdrop of 'local color'" (26). As our colleague Carolyn Mitchell has pointed out to us, the closing

credits sequence may be meant to evoke less the anodyne middle-class vacation spot than a sudden resurgence of an ambivalently invested exoticism. The logic of this reading would have Lecter's cannibalism finally finding its most "appropriate" environment; that is, his particular savagery would be linked, in accord with the drift of centuries-old racist associations, with the savage black "exotics" he strolls among. In the final scene, however, Lecter's white linen suit serves so forcefully to mark him off from the black figures in the street that it seems difficult to ignore the visual message of their separateness. It seems more apt to consider the sudden foregrounding of black people as figuring an essential victimhood, here inescapably associated with racial oppression and economic exploitation. As we have argued throughout, we continue to identify with this victimization, even if as only one side of an intractable ambivalence, the other side of which is embodied in Lecter.

28. In a longer essay, this would be the place to express whatever reservations we have about Žižek's own work, and in particular about his insistence, in many places, that to talk about the subject at all is perforce to talk about the Enlightenment subject, of which the postmodern subject is the secret truth rather than its negation or rupture—that, for example, "those who preach 'multicultural decenterment,' 'openness toward non-European cultures,' etc., thereby unknowingly affirm their 'Eurocentrism,' since what they demand is imaginable only within the 'European' horizon" (*Enjoy*, 185). In terms of the matter at hand, we want to characterize this as Žižek's overreadiness to remain within a rigid Cartesian dualism of "man" versus "nature"—one that has been subjected to devastating criticism in pragmatist critique and recent philosophy of science, among others—which Žižek does not so much rethink as simply *invert*.

From the point of view of the problem of species, what is wrong here is that the animal other, rather than enabling a decomposition of the terms of Enlightenment subjectivity as such, is in the curious position of "loser wins." If the animal other possesses any power to demystify the Cartesian "subject who knows," it does so only by virtue of how completely it merges with the realm of brute matter that must be repressed, with the obscene object, the stain, and so on. To the extent that the animal other is recognized as some sort of subjectivity—which is, we should note, undeniably the case in many disciplines outside psychoanalysis—it loses any critical potential. [I will revisit these points below in my discussion of Hemingway's later work.—C. W.]

29. Fredric Jameson, *Postmodernism, or The Cultural Logic of Late Capitalism* (Durham, N. C.: Duke University Press, 1991), 152. For more on "nominalism," see Jameson's chapter "Immanence and Nominalism in Postmodern Theory." See also his *Late Marxism: Adorno, or The Persistence of the Dialectic* (London: Verso, 1990), 157–61 passim.

Chapter 4

1. See recent studies such as Mark Spilka's *Hemingway's Quarrel with Androgyny* (Lincoln: University of Nebraska Press, 1990), Nancy R. Comley and Robert Scholes's *Hemingway's Genders: Rereading the Hemingway Text* (New Haven: Yale University Press, 1994), and Rose Marie Burwell's *Hemingway: The Postwar Years and the Posthumous Novels* (Cambridge: Cambridge University Press, 1996). Further references to all these works are given in the text.

2. Spilka calls *Eden* Hemingway's "most experimental and easily his most ambitious novel" (280)—a sentiment shared even by more skeptical reviewers such as E. L. Doctorow, whose famous piece in *New York Times Book Review* finds the novel an admirable failure in its attempt to treat new themes and problems: "That he would fail is almost not the point—but that he would have tried, which is the true bravery of a writer" ("Braver Than We Thought," reprinted in *Ernest Hemingway: Six Decades of Criticism,* ed. Linda W. Wagner [East Lansing: Michigan State University Press, 1987], 331).

3. Arnold E. Davidson and Cathy N. Davidson, "Decoding the Hemingway Hero in *The Sun Also Rises,*" in *New Essays on "The Sun Also Rises,"* ed. Linda Wagner-Martin (Cambridge: Cambridge University Press, 1987), 91. Further references are given in the text.

4. As one commentator has noticed of the moment on the Seine, "Geography has little to do with this. . . .Bill feels so good that he doesn't need a drink"—almost unheard-of for Gorton. "In fact, Jake and Bill are almost always in good spirits when together, either alone or with other male companions" (Scott Donaldson, "Humor in *The Sun Also Rises,*" in Wagner-Martin, *New Essays on "The Sun Also Rises,"* 37).

5. Ernest Hemingway, *A Moveable Feast* (1964, rpt. New York: Simon and Schuster, 1996), 190.

6. Jean-Paul Sartre, *Being and Nothingness,* trans. Hazel Barnes (New York: Simon and Schuster, 1956), 345–46. Further references are given in the text.

7. Stephen Melville, "In the Light of the Other," *Whitewalls* 23 (fall 1989): 13. Further references are given in the text.

8. To address Melville's very interesting discussion (23–24) of Jacques Lacan's discourse on the difference between humans and animals with regard to vision and mimicry in *The Four Fundamental Concepts of Psychoanalysis* would unfortunately take us too far afield of my topic. But Lacan's discussion should be consulted by anyone seriously interested in his relation to "carnophallogocentrism."

9. Slavoj Žižek, *The Sublime Object of Ideology* (London: Verso, 1989), 69. Further references are given in the text.

10. Walter Benn Michaels, *Our America: Nativism, Modernism, and Pluralism* (Durham, N. C.: Duke University Press, 1995), 13. Further references are given in the text.

11. For my purposes, the primacy of the ur-discourse of species would be the very point of Michaels's footnote to Marc Shell's discussion of the history of bullfighting in Shell's *Children of the Earth* (New York: Oxford University Press, 1993). The bullfight, Shell tells us, became a great national festival in Spain in the wake of the Christian reconquest and the expulsion of the Jews in 1492 and of the Muslims in 1502. "At the heart of this reconquest," Michaels tells us, "were the Statutes of the Purity of the Blood," which "changed the difference between Christians and non-Christians from a difference of religious practice into a difference of blood. . . . Shell identifies this transformation with what he regards as the underside of Christianity's universalist understanding of all humans as brothers—those who are not my brothers are not human" (Michaels, 163 n. 133).

12. Georges Bataille, *Theory of Religion,* trans. Robert Hurley (New York: Zone, 1989), 40. Further references are given in the text.

13. In a sense, of course, the Davidsons are right, for part of what is wrong with

heteronormative male sexuality by its own logic is that it instantiates a fatal dependence on women that is *spurious*—and indeed potentially injurious—to heteronormative *cultural* maleness as a homosocial practice.

14. For Spilka's discussion, see 225–27.

15. See also here Comley and Scholes's very interesting discussion of Hemingway's view of the decadence of bullfighting in relation to the other arts, especially
painting (on 118 ff.).

16. Max Eastman, "Bull in the Afternoon," in *Ernest Hemingway: The Man and His
Work*, ed. and intro. John K. M. McCaffery (New York: Avon, 1950), 57–58.

17. It also unmistakably connects Catherine Bourne to Brett Ashley in *The Sun
Also Rises*. When the narrator describes Catherine's dramatic entrance after having
her hair cut, we can't help being reminded of Brett and the nearly epithetical description we get time and again of her "wrinkling the corners of her eyes" (*Sun Also
Rises*, 78): "He heard her come into the cafe and say in her throaty voice, 'Hello dar-
ling.' She came quickly to the table and sat down and lifted her chin and looked at
him with the laughing eyes and the golden face with the tiny freckles. Her hair was
cropped short as a boy's. It was cut with no compromises" (14–15).

18. Here, of course, we recognize a characteristic Hemingway theme and, beyond that, a characteristic irony of modernist writing generally: that the only world
in which one can be happy is a world that is too "simple" to exist. This is the same
world we find in the "Bimini" section of another posthumous novel from Hemingway, *Islands in the Stream*; but in that world "there are no women to complicate the
lesson in how men must live" (Spilka, 262), as the painter Thomas Hudson spends
the days fishing and swimming with his male friends and his sons. The ending of that
novel, however, seems only to confirm that Hemingway thought that world too
"simple" as well. As he writes in a letter in November 1952 that is critical of the
readiness of his readers to buy into just such "simplicity," "[I am] not going to slant
my stuff for those high school kids who read OMS [*The Old Man and the Sea*] out
loud in class and write you identical, touching appreciations of it. [He then mentions
plans to eventually publish] the idyllic book about the Sea [that became *Islands*]
which we hope nobody will notice ends tragically. By that time maybe the younger
readers of OMS will be grown up enough to read the next two" (qtd. in Burwell,
133).

19. Interestingly enough, this links *Eden* to the "decadent" modernism of a
Baudelaire, and also to the "decadence" of the bullfight in *Death in the Afternoon* that
I have already discussed. To combine this aspect of the Hemingway text with his
modernist antiromanticism, I might put it this way: the Eden of iterative pleasures
may indeed be a too-"simple" place from which the human is barred, but the human
does not secure transcendence of it by obeying the injunctions and sacrificial
economies of humanism.

20. See Burwell, 110. As Burwell points out, while the novel offers the "red her-
ring" of Catherine's failure to conceive a child as the source of her frustrated creativity, "the manuscript makes it clear not only that artistic creativity is her highest
priority, but that she loathes children and that David suspects her failure to conceive
may be the result of *his* sterility" (111).

21. See Burwell, 214 n. 7, for an overview of the various critical responses to the
elephant hunting narrative.

22. See, for example, Burwell, 120–21, 125, and Carl Eby, "'Come Back to the Beach Ag'in David Honey!' Hemingway's Fetishization of Race in *The Garden of Eden* Manuscripts," *Hemingway Review* 14, 2 (spring 1995): 109.

23. See also Comley and Scholes, 95.

24. For examples of this view of the symbolic significance of the elephant, see James Hillman, "The Elephant in *The Garden of Eden*," *Spring: A Journal of Archetype and Culture* 50 (1990): 93–115, and Eby, "'Come Back to the Beach.'"

25. Gilles Deleuze and Félix Guattari, *A Thousand Plateaus: Capitalism and Schizophrenia*, trans. Brian Massumi (Minneapolis: University of Minnesota Press, 1987), 240–41.

26. Slavoj Žižek, *Enjoy Your Symptom! Jacques Lacan in Hollywood and Out* (New York: Routledge, 1992), 117. Further references are given in the text.

27. Judith Roof, *Come as You Are: Sexuality and Narrative* (New York: Columbia University Press, 1996).

28. This "mastery" cannot be entirely laid at the feet of Scribner's editing of the manuscript, because it is there all along in the subtle but important presence of the *narrator*, who, as an avatar of the father, structures the novel and makes available to us everything we know about David, including his childhood memory. This avuncular narrative presence recasts the ambivalent David in its own image in a relationship strikingly similar to what we find between the narrator and the young Hemingway in Paris in the contemporaneous *A Moveable Feast*. This is the narrator who intervenes at key moments to reframe and disavow David's cross-gender and cross-species identification in light of an Oedipal knowledge not yet fully earned by the young writer. For example, when David looks in the mirror in the passage noted earlier and says, "You like it. Remember that. . . .You know exactly how you look now and how you are," the narrator in hindsight intones, "Of course he did not know exactly how he was" (85). Or again, this time on the site of not cross-gender but cross-species identification: "Many times during the day he had wished that he had never betrayed the elephant and in the afternoon he remembered wishing that he had never seen him." To which the narrator responds, "Awake in the moonlight he knew that was not true" (174). Or again, later in the novel, when David remembers, "The elephant was his hero now as his father had been for a long time," such knowledge is immediately disavowed from the vantage of David-the-adult, which is itself enframed within the knowledge of the narrator-as-even-more-adult: "It was a very young boy's story, he knew, when he had finished it" (201).

29. Diana Fuss, *Human, All Too Human* (New York: Routledge, 1996), 2–3.

30. And this is quite literally so. As I have noted in earlier chapters, Freud wrote to Wilhelm Fliess in 1897—in terms that anticipate by thirty-plus years the formulations of *Civilization and Its Discontents:* "Perversions regularly lead to zoophilia and have an animal character. They are explained not by the functioning of erogenous zones that later have been abandoned, but by the effect of erogenous *sensations* that later lose their force. In this connection one recalls that the principal sense in animals (for sexuality as well) is that of smell, which has been reduced in human beings. As long as smell (or taste) is dominant, urine, feces, and the whole surface of the body, also blood, have a sexually exciting effect" (*The Complete Letters of Sigmund Freud to Wilhelm Fliess, 1887–1904*, trans. and ed. Jeffrey Moussaieff Masson [Cambridge Belknap Press of Harvard University Press, 1985], 279).

31. Indeed, Žižek's whole polemical and theoretical distinction between Lacan's project and that of poststructuralism is based on this ontological positivity, as he notes in the introduction to *Tarrying with the Negative: Kant, Hegel, and the Critique of Ideology* (Durham, N. C.: Duke University Press, 1993), 3–4.

32. Judith Butler, *Bodies That Matter: On the Discursive Limits of "Sex"* (New York: Routledge, 1993), 218–19.

33. Diana Fuss, *Identification Papers* (New York: Routledge, 1995), 2. Further references are given in the text.

34. I borrow this notion of "the lure" from Rey Chow's "The Dream of a Butterfly," in Fuss, *Human, All Too Human*, 69–70.

35. Here Fuss's discussion of the relation between identification and desire in Freud is very much to the point, particularly regarding what Catherine wants—that is, Does Catherine desire those of her same sex and thus identify with David the male, a fantasy of male prowess and power that David (who does not possess it) thus identifies with via Catherine? To put the question this way, however, is to notice that the Freudian distinction itself—that identification is "the wish to be the other" and desire is "the wish to have the other"—is, as Fuss puts it, "a precarious one at best, its epistemological validity seriously open to question" (*Identification Papers*, 11).

36. Jacques Derrida, "The Animal That Therefore I Am (More to Follow)," trans. David Wills, 47 (unpublished manuscript; see chap. 2, n. 22 above).

37. Donna J. Haraway, *Simians, Cyborgs, and Women: The Reinvention of Nature* (New York: Routledge, 1991), 192. Further references are given in the text.

38. Brian Massumi, *A User's Guide to Capitalism and Schizophrenia: Deviations from Deleuze and Guattari* (Cambridge: MIT Press, 1992), 83, 84. Further references are given in the text.

39. Harriet Ritvo, "Barring the Cross: Miscegenation and Purity in Eighteenth- and Nineteenth-Century Britain," in Fuss, *Human, All Too Human*, 52.

40. Bell Hooks, "Eating the Other: Desire and Resistance," in her *Black Looks: Race and Representation* (Boston: South End Press, 1992), 21–22. Further references are given in the text.

41. Toni Morrison, *Playing in the Dark: Whiteness and the Literary Imagination* (New York: Random House, 1992), 90. Further references are given in the text.

42. Etienne Balibar, "Racism and Nationalism," in *Race, Nation, Class: Ambiguous Identities*, ed. Etienne Balibar and Immanuel Wallerstein, trans. Chris Turner (London: Verso, 1991), 56, 57.

43. These facts about Leda and the Swan in the Prado are discussed in Comley and Scholes, 95–96. See here, once again, Catherine's association with ivory and, further afield, Thomas Hudson's relations with his cats in the contemporaneous *Islands in the Stream* (New York: Scribner's, 1970), 204–5, 212–13, 221–22, 238–39. These sections of *Islands* are interesting in that they threaten to push beyond the Oedipal recontainment of the pet economy (noted by Deleuze and Guattari) in their explicit eroticization of relations across species lines. See also here Spilka, 263, who associates the cats in *Islands* (one of whom is named "Litless") with Catherine Bourne.

Chapter 5

1. Gilles Deleuze and Félix Guattari, *A Thousand Plateaus: Capitalism and Schizophrenia*, trans. Brian Massumi (Minneapolis: University of Minnesota Press,

1987). Further references are given in the text. It is a critique that, in its extremity, raises the question whether *even* the figure of the "stand" or "stance" as I use it in my introduction, following Cavell's work, remains too tied to identity and the gaze of the human to be of much use. See their critique of the regime of "faciality" in *A Thousand Plateaus*, a regime that is for them more foundational than the regime of the look (171). As Brian Massumi characterizes it, "The 'face' in question . . . is less a particular body part than the abstract outline of a libidinally invested categorical grid applied to bodies (it is the 'diagram' of the mother's breast and/or face abstracted from the maternal body without organs and set to work by the socius toward patriarchal ends)" (172 n. 54). "Faciality" thus refers essentially to a fetishized localization of desire whose aim is fixity and identity, and whose apotheosis is the Oedipal regime. This is why "the form of subjectivity," as Deleuze and Guattari put it, "whether consciousness or passion, would remain absolutely empty if faces did not form loci of resonance that select the sensed or mental reality and make it conform in advance to a dominant reality. The face itself is a redundancy" (168). And thus "if human beings have a destiny, it is rather to escape the face, to dismantle the face and facializations, to become imperceptible, to become clandestine, not by returning to animality, nor even by returning to the head, but by quite spiritual and special becomings-animal . . . that make *faciality traits* themselves finally elude the organization of the face—freckles dashing toward the horizon, hair carried off by the wind" (171). This critique of faciality, of course, has very direct implications for our reading of Lévinas's ethics in relation to the face, as discussed in chapter 2. See Brian Massumi, *A User's Guide to Capitalism and Schizophrenia: Deviations from Deleuze and Guattari* (Cambridge: MIT Press, 1992).

2. A question raised by Deleuze and Guattari's reorientation of our discussion of animality in the direction of multiplicity is raised in a different register by Donna Haraway's work: As for the ethical status of nonhuman others, do *animal* nonhuman others have priority? Not according to Haraway, whose cyborg would take its place alongside the chimpanzees of animal rights philosophy and the wolf packs of *A Thousand Plateaus*. Those who find Haraway's assertion counterintuitive or implausible should consult the wonderful episode of *Star Trek: The Next Generation*, in which the android Commander Data argues in court—successfully—for his right not to be disassembled. Whether this explains or mercilessly ironizes the fact that Data is the crew's leading anthrophile, we will have to leave for another discussion!

3. Michael Crichton, *Congo* (New York: Ballantine Books, 1980). Further references are given in the text.

4. See Fredric Jameson's characterization in *Postmodernism, or The Cultural Logic of Late Capitalism* (Durham, N. C.: Duke University Press, 1991), 1–54.

5. For an overview of Sue Savage-Rumbaugh's work with Kanzi, and of ape-language experiments in general, see her essay in *The Great Ape Project: Equality beyond Humanity*, ed. Paola Cavalieri and Peter Singer (New York: St. Martin's Press, 1993).

6. Toni Morrison, *Playing in the Dark: Whiteness and the Literary Imagination* (New York: Random House, 1992), 45, 39. I appreciate comments on the novel as a kind of racial allegory that were suggested by Joanne Wood at an earlier stage of this chapter.

7. This makes the novel's caricature of the animal rights movement and its attacks on "Elliot and his Nazi staff" (38) early in the story all the more curious.

8. Sigmund Freud, *Civilization and Its Discontents*, trans. and ed. James Strachey (New York: Norton, 1961), 52.

9. This is the point that is missed, I think, in the recent reception of the novel in a prominent animal rights publication, *Animal People*. "The hero of *Congo* is Amy," the reviewer writes, "[who] seems to be the only character possessing sensitivity, wit, insight, and any true link to her surroundings. . . . Amy alone can feel the sense of the jungle, its nature, and its agenda." This reading of Amy—which is correct, as far as it goes—misses two crucial points: first, within the discourse that governs the universe of the novel, this is not necessarily good news. Yes, one is tempted to say, she does—*exactly like a child*. Second, as much might be said, after all, of Amy's other in the bifurcated animal category, the gray gorillas, who are best described in the diametrically opposed terms of Deleuze and Guattari's notion of "multiplicity" and the "pack" (or in this case the troop). See Pamela June Kemp, review of *Congo, Animal People* 4, 8 (October 1995): 22.

10. Olivier Richon, "The Hunt," *Public* 6 (1992): 89. In this light, that the gray gorillas use stone tools in their attacks is less a questioning of the species barrier than a confirmation of it as theorized by Georges Bataille's discussion in *Theory of Religion*, which I examined earlier in the book. For the tool, Bataille writes, is not a reliable sign of the distinctly human, because the meaning of the tool remains subordinated to "utility"—to function rather than contemplation—and thus remains tied to the world of the object, the world of "immediacy" in which the animal remains locked, and from which the human distances itself via its ability to create purely abstract, symbolic meaning in art and ritual. Georges Bataille, *Theory of Religion*, trans. Robert Hurley (New York: Zone Books, 1992), 36.

11. Slavoj Žižek, *Enjoy Your Symptom! Jacques Lacan in Hollywood and Out* (New York: Routledge, 1992), 22.

12. Quoted in Fredric Jameson, *Late Marxism: Adorno, or The Persistence of the Dialectic* (London: Verso: 1990), 214; emphasis mine. Žižek's assertion that the idea of life is alien to the symbolic order is a strong echo of Horkheimer and Adorno's critique of "administered society" and its view of nature as merely a fungible resource for calculated control and exploitation. Likewise, the "multiplicity" invoked by Deleuze and Guattari is a strong echo of Adorno's assertion that the aim of nonidentity theory or negative dialectics is to restore the other—whether the social other or nature in relation to society—to its proper status as "the preponderance of the object" that has been subsumed, in Enlightenment, under identity, the concept, and reification. For amplification of this latter point, see Fredric Jameson's masterful discussion in *Late Marxism*, esp. 94–110, 212–19.

13. Donna J. Haraway, "Situated Knowledges: The Science Question in Feminism and the Privilege of Partial Perspective," in her *Simians, Cyborgs, and Women: The Reinvention of Nature* (New York: Routledge, 1991), 189.

14. Philippe Lacoue-Labarthe, "Typography," in his *Typography: Mimesis, Philosophy, Politics* (Cambridge: Harvard University Press, 1989), 116.

15. Quoted in Michael Taussig, *Mimesis and Alterity: A Particular History of the Senses* (New York: Routledge, 1993), 19. Further references are given in the text.

16. Homi K. Bhabha, *The Location of Culture* (London: Routledge, 1994), 86. Further references are given in the text.

17. Donna Haraway, "The Biological Enterprise: Sex, Mind and Profit from Human Engineering to Sociobiology," in her *Simians, Cyborgs, and Women*, 44.

18. Fredric Jameson, "Periodizing the Sixties," in his *The Ideologies of Theory: Essays 1971–1986*, vol. 2, *The Syntax of History* (Minneapolis: University of Minnesota Press, 1988), 184.

19. I am thinking specifically of Horkheimer and Adorno's critique of Marxism as itself representative of the Enlightenment *episteme* that would, as Adorno put it, "turn all of Nature into a giant workhouse," treating Nature and the object world as mere fungible resources for exploitation and control. This is directly related, of course, to Adorno's critique of dialectic and its fetishizing of identity and the concept, and its inability to account for what he calls "the preponderance of the object." See, for a fuller discussion, Cary Wolfe, "Nature as Critical Concept: Kenneth Burke, the Frankfurt School, and 'Metabiology,'" *Cultural Critique* 18 (1991): 65–96.

Conclusion

1. Richard Rorty, *Objectivity, Relativism, and Truth*, Philosophical Papers 1 (Cambridge: Cambridge University Press, 1991), 110. Further references are given in the text.

2. Paola Cavalieri and Peter Singer, Preface to *The Great Ape Project: Equality beyond Humanity*, ed. Paola Cavalieri and Peter Singer (New York: St. Martin's Press, 1993), 1.

3. Cary Wolfe, *Critical Environments: Postmodern Theory and the Pragmatics of the "Outside"* (Minneapolis: University of Minnesota Press, 1998).

4. Zygmunt Bauman, *Postmodern Ethics* (Oxford: Basil Blackwell, 1993), 1, 21. Further references are given in the text.

5. On this point, Bauman succinctly notes, "'I am ready to die for the Other' is a moral statement; 'He should be ready to die for me' is, blatantly, not" (51).

6. John Llewelyn, "Am I Obsessed by Bobby? (Humanism of the Other Animal)," in *Re-reading Levinas*, ed. Robert Bernasconi and Simon Critchley (Bloomington: Indiana University Press, 1991), 236.

7. Richard Beardsworth, *Derrida and the Political* (London: Routledge, 1996), 130. Further references are given in the text.

8. Marc-Alain Ouaknin, *Méditations érotiques* (Paris: Balland, 1992), 129, qtd. in Bauman, 84–85.

9. This analysis amplifies considerably, I think, the critique of Homi Bhabha's "cultural translation" that I attempted to elaborate in chapter 5. For Bhabha, "what is in modernity more than modernity is this signifying 'cut' or temporal break" that is "in" modernity very much along the lines that the postmodern, for Lyotard, inhabits the modern as its repressed other, which in turn enables Lyotard's famous formulation that the postmodern actually *precedes* the modern that represses these in the name of identity and its privileged terms (Oedipalism, capital, the metaphysical concept). We could therefore say—to return to Bhabha—that while this "cut" or "break" does indeed open up a space for the "active cultural translation" that is central to Bhabha's critique, it *does not require it* (since that space is not *for* the subject of modernity, any more than the inhumanity of time is "for" the subject or the other— hence the relevance of what I called, in the previous chapter, the "slowness" of the animal). Indeed, one might well argue—following Derrida and Lévinas on the im-

portance of "passivity"—that the "rush" to valorize "active cultural translation" in Bhabha only threatens to reinstate the temporal structure of "for the subject" that the "muteness" and "slowness" of the animal other would keep open.

10. For an overview of Luhmann's work, see Wolfe, *Critical Environments*, esp. 64–84 and 117–25. See also William Rasch's excellent *Niklas Luhmann's Modernity: The Paradoxes of Differentiation* (Stanford: Stanford University Press, 2000).

11. William Rasch, "Immanent Systems, Transcendental Temptations, and the Limits of Ethics," in *Observing Complexity: Systems Theory and Postmodernity*, ed. William Rasch and Cary Wolfe (Minneapolis: University of Minnesota Press, 2000), 91. Further references are given in the text.

12. Here it worth noting the irony that Bauman has done extensive work on the Holocaust. For him that historical event would no doubt illustrate the insufficient presence of the moral in relation to society and its function systems, particularly the political system. For Luhmann it would illustrate precisely the opposite: what happens when the autonomy of the function systems is broken down by the "parasitical" moral code, so that properly political distinctions between government and opposition are rewritten in moral terms of good and evil. On the terrain of science, one can observe the same sort of parasitical reinstatement of the moral code, as racial and ethnic distinctions between different types (Aryan and Jew, most obviously) are re-coded similarly in terms of good and evil.

13. As Rasch points out, "If we remain within the immanence of systems that Luhmann not only advocates but sees as inescapable, we are left with this paradox. Ethics emerges as the by-product of a system's attempt to preserve its own reproduction from the ravagings of moral infection. The only moral preselection said to be ethically permissible is the preselection that guarantees the freedom of selection" (94). Thus, "What presents itself, in Luhmann, as descriptive of modernity also takes on the force of a prescriptive. The description of modernity as differentiated needs to be read as both an empirical fact—'differentiation exists'—and as an imperative—'differentiation ought to (continue to) exist'" (90). Hence what I have called Luhmann's "liberal utopianism," which behaves *as if* the full autopoiesis of horizontally distributed function systems were not constrained or compromised by the vertical dominance of power, capital, and so on. See Wolfe, *Critical Environments*, 73–78, 145–50.

14. Specifically in Rorty, *Objectivity, Relativism, and Truth*.

15. See Wolfe, *Critical Environments*, 14–15.

16. Niklas Luhmann, "The Cognitive Program of Constructivism and a Reality That Remains Unknown," in *Selforganization: Portrait of a Scientific Revolution*, ed. Wolfgang Krohn, Günter Küppers, and Helga Nowotny (Dordrecht: Kluwer, 1990), 76; my emphasis.

17. Niklas Luhmann, *Ecological Communication*, trans. John Bednarz Jr. (Chicago: University of Chicago Press, 1989), 22–23. As for the "outside" of the system, the environment, Luhmann writes that it "is the total horizon of information processing that refers beyond the system. . . .[T]he system's environment has no boundaries nor needs any. Presenting itself as horizon, it is the system-internal correlate of all references that extend beyond the system. . . .The horizon always recedes when it is approached, but only in accordance with the system's own operations. It can never be pushed through or transcended because it is not a boundary. It accompanies every

system operation when this refers to something outside the system. As a horizon, it is the possible object of intentions and communication; but only in so far as the system can present the environment to itself as a unity—and this requires that it can differentiate itself as a unity from it" (22).

18. Niklas Luhmann, "The Paradox of Observing Systems," *Cultural Critique* 31 (fall 1995): 44. Further references are given in the text.

19. As Rasch puts it, "What presents itself, in Luhmann, as descriptive of modernity also takes on the force of a prescriptive. The description of modernity as differentiated needs to be read as both an empirical fact—'differentiation exists'—and as an imperative—'differentiation ought to (continue to) exist'" (90). The constitutively paradoxical act of observation, in generating the necessity of its unfolding in another observation, a second-order observation, thus generates a form of pluralism that in Luhmann takes on not only a descriptive (i.e., scientific) function but also an ethical one.

20. I develop this point with reference to Rorty in some detail in Wolfe, *Critical Environments*, 18–22, 65, 70–75, 141–43.